BUILDING & FLYING THE MITCHELL WING

Other TAB Books by the author:

No. 2264 *How To Fly Helicopters*

This book was written with the support and expert advice of those at Mitchell Aircraft Corporation, namely Don Mitchell, James Meade, Rusty Antles and "Ruthie," without whose help this book would have been an impossibility.

BUILDING & FLYING THE MITCHELL WING

BY LARRY COLLIER

FIRST EDITION

FIRST PRINTING

Printed in the United States of America

Library of Congress Cataloging in Publication Data

Collier, Larry.
Building & flying the Mitchell wing.

Includes index.
"TAB book #2302."
1. Airplanes, Home-built. I. Title. II. Title:
Building and flying the Mitchell wing.
TL671.2.C56 629.133'343 81-9251
ISBN 0-8306-2302-7 (pbk.) AACR2

Contents

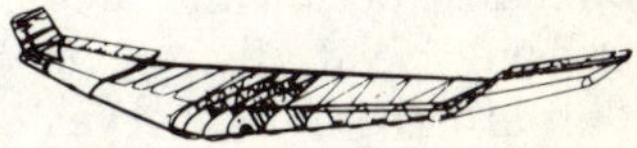

Foreword

Ever since the big gas crunch hit general aviation, as we entered the decade of the 1980s, sport flyers have been faced with a terrible dilemma — how to continue flying without going for broke?

Answers have come from many pilots themselves, and from such farsighted manufacturers as the Mitchell Aircraft Corporation in Porterville, California, whose fine products are described by veteran pilot Larry Collier in this exciting book.

The problem, of course, has been to develop fuel-efficient aircraft, the same motivation that led the late, great Bill Lear to design a thrilling new business aircraft, the Lear Fan, which is five times more fuel-efficient than the Lear Jet, and nearly as fast (twin turbos drive a single pusher prop).

At first, the ultralight movement got swinging with powered adaptions of early hang gliders that appeared at the start of the last decade; now the 1980s promises to be one of the wildest decades in the history of sport flying.

Much effort has been expended in development of new powerplants, or modifications of existing two-cycle engines, such as the McCulloch 101 chain saw machine. It operates at very high rpm, and yet is extremely reliable.

Other designers have gone in for novel aerodynamic configurations to attain the best performance at low Reynolds numbers, a sort of offshoot of model-plane technology. High aspect ratio wings . . . canard elevator surfaces. . . low profile tandem two-seaters . . . flying wings . . . all have been likely candidates for the sport plane designs of tomorrow.

In the latter category, of course, we find the Mitchell Wing B-10 and its big brother, the Mitchell Super Wing U-2. Both craft were designed by Don Mitchell, whose soaring experience goes back 40 years, to the time of the Northrop YB-49 flying wing bomber.

Pusher powerplants have recently come into vogue in the ultralight movement, not only in the Mitchell line but in many other craft that permit the pilot to sit out front and bite bugs at low cruising speeds, flying over the lovely countryside in the manner of the old barnstormers of yesteryear.

Water flying has become more and more popular as some ultralight drivers have added floats to their tiny machines, permitting long water runs and the potential of visiting many back country lakes and rivers unavailable to airmen flying more powerful craft that can't make it into puddle country.

All of this effort, of course, is under the control of the government — FAA regs still apply, although they have been encouraging the younger birdmen to go ahead and fly (off airways, naturally) on their own, so long as they don't hurt anybody else.

While no hard and fast definitions have as yet appeared to say what is and what is not a certifiable aircraft, the accepted criteria now seems to be whether it is capable of foot launch and landing, dispensing with conventional landing gear found on all conventional land planes.

The Mitchell people wisely elected to categorize their U-2 and P-38 as certifiable aircraft, which means they carry a government license number, and the pilot must have at least a student's pilot license. Further, they recommend all novice pilots to go and take dual instruction in conventional trainers prior to soloing the single-place flying wings, a wise idea.

As a certificated flight instructor who began teaching airmen in the Royal Air Force during World War II, I have to agree that the proper approach to ultralight flying is the cautious route, because danger lurks in the sky always for those who disobey the laws of nature.

But the wholesome thing about the movement today is that so far designers, pilots, homebuilders, the FAA, etc., all are working toward the same goal — safe flight. That doesn't, by any means, take the fun out of flying; it simply insures that we all get the most enjoyment from the sky in a safe and sane manner.

There have been fatalities, unfortunately, and most of them have been attributable to pilot error. Others have been linked to faulty designs that produce such things as uncontrollable pitch instability.

The homebuilt aircraft movement, sadly, doesn't have the means of testing models of their ships in expensive government research wind tunnels. But some, like Burt Rutan (designer of the composite construction VariEze, VariViggen, Defiant and Quickie) ingeniously developed cartop model mechanisms that serve as "wind tunnels" when driven 'like sixty' down country roads.

Such ingenuity is part of the fun of the movement, for whenever there is a challenge, there is someone who figures a way to meet it. Getting the act together is not as simple, for the laws of aerodynamics are not well understood by the amateur. A domino effect exists: one design error can result in a whole series of follow-on problems, unless corrected quickly.

Thus we can appreciate what Larry Collier is trying to tell us — go to the master, young man, and listen to what he says. In this case, the people who developed the Mitchell Wing ultralights are leaders in the field. Like any other aircraft, they will be improved over the years, a method of growth we hope will be found throughout the homebuilt movement, because you just cannot stand still in this business — you go forward or fall behind.

Hopefully, there is a bright future ahead for all of us in low-cost flying, as we accumulate innovative ideas to make the old tried-and-true designs work better for us.

Don Dwiggins
Editor
Homebuilt Aircraft

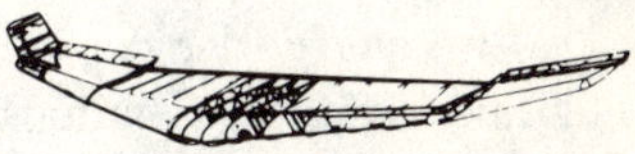

Amateur Builder's Information and Guidelines 1

To those who have never tried their hand at homebuilding, the thought of aircraft construction must surely bring forth nightmarish thoughts of blueprints and plans by the ream, as well as sophisticated tooling that only experts are capable of fathoming. However, such ideas are soon swept into an out-of-the-way corner of the mind with the Mitchell Wing.

Full sets of plans are drawn in a fashion that is easy to understand and even simpler to use. They are mostly full-sized, and include an instructional manual consisting of material lists for each stage of building and photos that lead you step-by-step to the finished product.

This book can by no means take the place of either a set of plans or a kit. However, it can be considered one of the most in-depth informational packages that you'll ever read. And, used in conjunction with plans or kit, it will brush you through areas where builders normally get nailed.

Once you've committed yourself to a Mitchell Wing and purchased a set of plans or kit, a whole new world opens to you as a fun-flyer. Mitchell follows through, working with you during, and after building your ultra-light. A one-year membership comes with the package and includes newsletters, invitations to building seminars, modification updates, special events, get-togethers and fly-ins.

When a set of plans or a kit is purchased, it receives a serial number, kept on file at Mitchell. This, in essence, assures you of up-date and modifications pertaining to your particular purchase, when they occur. This service is free with purchase for the first year. Thereafter, there's a nominal yearly fee.

For more information, write or call: Mitchell Aircraft Corporation, 1900 S. Newcomb, Porterville, CA 93257, telephone (209) 781-8100.

You take it from here. But no matter which way you decide to go with such an undertaking, I think that you'll be pleasantly surprised.

After you've decided to build the Mitchell Wing, you'll receive help and further information by contacting the local FAA Engineering & Manufacturing District Office and letting them know of the make and model of aircraft you're going into. Give them your name, address, and telephone numbers, along with detailed directions on how to get to the inspection site.

The following two inspections can add health and longevity to both your machine and yourself. They are recommended in your interests.

Step One: In-Process Inspection

These "in-process" inspections pertain to all major components, such as fuselage, wings, and control surfaces. These should be ready for closing at the time of inspection. You should be ready and able to satisfactorily demonstrate attaching fit of engine mount, landing gear, and installation of control systems—including routing, clearances and control travel stops.

☐ During related stages of actual aircraft construction, it's in your best interest to prepare some samples of the most difficult welding and wood gluing joints for later examination. Use leftover pieces of representative material for samples. Keep a record of all tests you conduct, listing material used and test results.

☐ Note minor changes to the original design directly on the drawings, and indicate same to inspector. If you've made major changes to the original design, submit them to Mitchell Aircraft Corporation or an aeronautical engineer for analysis and evaluation. Also send a copy to the inspector prior to closing the area involved.

☐ Keep an aircraft logbook available for inspection acceptance of components to be closed.

☐ When your assemblies are ready for inspection, contact the inspector for a scheduled visit. During this inspection expect a review of drawings. procedures and workmanship.

☐ The following regulations are noteworthy, and manuals containing valuable information can be ordered through the

Superintendent of Documents, U.S. Government Printing Office, Washington, DC 20402:

FAR 45 — Identification and Registration Marking

FAR 47 — Aircraft Registration

FAR 91 — General Operating and Flight Rules

AC43.13-1A, Stock #5011-0058 — Acceptable Methods, Techniques, and Practices, Aircraft Inspection Repair

AC48.13-2A, Stock #050-007-00407-9 — Acceptable Methods, Techniques and Practices, Aircraft Alteration.

AC65-9A, Stock #050-007-00379-0 — Airframe and Powerplant Mechanics General Handbook

AC20-106, Stock #050-007-00449-4 — Aircraft Inspection for the General Aviation Aircraft Owner

Note: The Experimental Aircraft Association (EAA), Box 229, Hales Corners, WI 53130, has a complete line of manuals, data plate and logbook information for amateur builders. Membership in this worldwide organization, as well as a local chapter, is highly recommended. Availing yourself of one of their EAA Designee services will keep everything in line and you on the right track.

Step Two: Registration

You'll need an identification "N" number for your Wing prior to final inspection. This "N" number is obtained from: FAA, Aircraft Registration Branch, P.O. Box 25504, Oklahoma City, OK 73125, by submitting AC Form 8050-88, "Affidavit of Ownership for Amateur-Built Aircraft" — or a letter of request stating your name, make, model, serial number (selected by you), class, type of engine installed, number of engines, manufacturer, model, serial number of each engine(s) installed, land or water operation, number of seats, and a notarized statement that the aircraft is built from parts by you, and that you're the owner. It's further recommended that the "N" number be requested approximately 60 days before you're ready to paint the "N" number on your Wing. If you'll accept any number selected by the Aircraft Registration Branch, there's no charge. However; if you apply for a special "N" number, there's a $10 charge. Send the money with your "N" number request. Don't send the application for aircraft registration, AC Form 8050-1, at this time, since this would only cause a delay in the issuance of your "N" number.

The application for registration, AC Form 8050-1, along with a $5 registration fee needs to be sent to the same address, but only a few days prior to the FAA's final visit for airworthiness certification. The pink copy of the application is yours to keep as authorization to fly for 90 days, or until registration is returned from Oklahoma City, whichever comes first.

Step Three: Airworthiness Certification

After the aircraft is completed and all ground testing, including fast taxi test, are conducted to determine controllability, the FAA should be contacted for a final inspection for airworthiness certification.

At this time, you must submit to them an application for airworthiness certificate, FAA Form 8130-6, filled out in duplicate (front of form only) and a letter indicating the desired flight test area by longitude and latitude cordinates—or by a radius from a reference point—in each case including the base airport, or a corridor to and from the base of operation.

Step Four: Final Inspection

At the time of final inspection, all inspection covers and fairings covering wing attachment to fuselage need to be removed and the engine cowling (not on the Mitchell) opened. The aircraft actual weight and balance data must be available for review by the inspector. The following are a few of the items the FAA inspector will be checking on:

☐ Proper control rigging and full travel.

☐ Fuel shut-off and selector valves easily accessible in flight position, properly identified, and easy to operate.

☐ Throttle movement is smooth and positive, with no lag or spring-back.

☐ The installation of shoulder harnesses is highly recommended. Seat belts and/or shoulder harnesses securely fastened to the airframe.

☐ Aircraft equipped with retractable landing gear will require a demonstration of the gear operation at final inspection with the aircraft jacked up in such a manner as to simulate flight load on the wings.

☐ Trim system operating smoothly throughout its range, if applicable.

☐ Instruments having range markings on the glass have a slip mark on the glass and case as an indicator of glass rotation due to panel vibration.

It should be understood that the FAA, besides being an overseer, is also a help and wealth of information for the person who is tackling any form of aviation. Check in with them. They can give you what you need—communicate.

Engines and Propellers 2

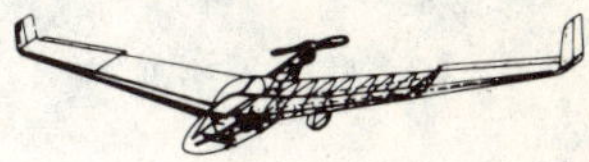

McCulloch

One of the most basic, and best, all-round engines for modification, thrust/power vs. rpm, and life-span is the McCulloch 125cc (Figs. 2-1 and 2-2). Even though it's just a raw-boned chain saw engine, as the saying goes, "You treat it right, and it'll treat you right."

One of the only problem areas encountered with the McCulloch is in trying to pull *too much* power from it for *too long* a time period. This, however, can be said of most any engine. Engine temperatures rise, and pistons seize. It's a costly mistake, especially if you don't do the overhaul yourself.

The Mitchell ships utilize the McCulloch very successfully. Best overall results are in conjunction with a 3:1 ratio reduction system. This means for every three engine revolutions, the prop turns only once. Given 9000 rpm, say on takeoff and climb, the propeller will only be revving 3000 rpm. Noise level is kept to a minimum with this system.

Propellers used with the McCulloch are either a 48″×22″-pitch or 48″×24″-pitch. The only difference is that the former is considered a takeoff and climb prop, and the latter is one used for cruise. The 48″×22″-pitch is preferred.

Turning full rpm, using the 3:1 reduction system and 48″×22″-pitch propeller, the McCulloch develops about 10 hp, giving the propeller over 100 pounds thrust. All of this while sipping slightly more than 1 gph of fuel/oil mix.

Fig. 2-1. McCulloch 125cc on B-10 cage.

Zenoah

For those high-altituders, waterwingers, snowskiers and real hefties needing more power, Mitchell has bench and flight tested other engine modifications. One of these is the Zenoah 250cc.

This Japanese lightweight hooks on ready-to-fly at about 45 pounds, including reduction system and propeller. It's recommended that the Zenoah be used with the U-2 only. This recommendation is made only because of weight distribution. The Zenoah 250cc is 15 pounds heavier than the McCulloch, and it's sitting in the extreme aft position. This simply means that with the B-10, a lot more weight must be added to the front of the ship to offset it, or its moment/arm must be cut, bringing it closer to datum line.

In conjunction with a 2:1 reduction system, there are a number of wooden, handcarved propellers that fit into its pusher mode. The 45″×20″-pitch climb prop produces just about 150 pounds of thrust with the engine turning 7000 rpm. That's equal to approximately 20 horses. At 5500 rpm, the Zenoah is pumping out an easy 16 hp.

Using the 7000 rpm power setting for max climb, a 45″×20″-pitch climb prop, and a 2:1 reduction system, the U-2 will climb out at 500 fpm plus. Cruise will top out way above the 60 mph mark. It'll get off the ground in 200 feet, and land within 150. With the cruise prop, same conditions, takeoff is increased to something around 300 feet—no change in landing distance.

Honda

Another popular engine is the Honda 250cc, also tested with the 2:1 reduction system, although other reductions are available. This is the same basic engine designed for off-road vehicles, so it has proven to be rough, tough and able to take abuse. Large cooling fins dissipate heat readily, giving you a closer engine temperature control for maximum efficiency (Fig. 2-3).

Using a 48″×26″-pitch propeller, the system develops more than 130 pounds thrust, with the engine turning the pusher at 3400 rpm. Climbout, in this configuration, is near 500 fpm, with normal cruise in the 50-55 mph range. This is, of course, utilizing the climb prop.

Equipped with a cruise prop, expect high top-end speeds, but takeoff and climb figures are lost — long ground run and slower

Fig. 2-2. McCulloch 125cc on U-2.

Fig. 2-3. Honda 250cc on B-10 cage.

climb footage. As with the 48″×26″-pitch, though, climb props give excellent takeoff and climb, but lower top-end speeds. It's up to you.

In the ultralight field, most builders seem to prefer a climb propeller, or one close to it. The reason: it's a fun-flying machine, not based on high speeds, but rather, on getting off the ground and climbing to altitude, where soaring, gliding and thermaling can be enjoyed with the engine off or at idle.

The Honda 250cc is rated at 18-20 hp at 6800 rpm, carrying the Super Wing to top speeds nearing 70 mph. Climb is accomplished at 6000-6200 rpm, and cruise at 5000, Weigh-in for a ready-to-fly system is 52-55 pounds—a little heftier than either the McCulloch or the Zenoah. Fuel consumption is estimated at roughly 1.5 gph.

Add-Ons

It's recommended that Mitchell Wing engines be equipped with tuned exhausts. Take the McCulloch, as an example. You'll appreciate the difference when the engine reaches 7000 rpm and it comes into play. It's almost as if a switch has been thrown. A tuned exhaust lets the engine work at max efficiency, especially when hitting the 8000 rpm mark and above, while drastically reducing engine heat.

With the optional engines burning more fuel than the basic McCulloch, there's a clear-cut need for larger fuel tanks than the standard 2-gallon. Some builders have modified their craft to take four and six gallon tanks. This, of course, will greatly extend the range/time in the air for those not so willing to let nature provide the power.

As with most homebuilts, trying the unusual, on-your-own experiments with different equipment and modifications can sometimes wreak havoc on both the machine and its pilot's health. Always check with knowledgeable personnel before trying any such venture. Use Mitchell's motto when approaching the unknown: "Use only tried and true methods and materials." In other words, let someone else take the worry out of your flying.

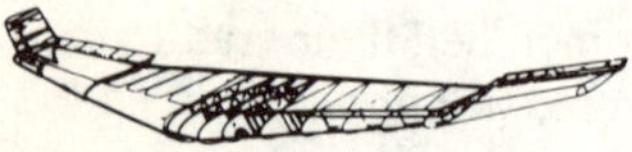

3 Bonding

Before beginning either the B-10 or the U-2, you need to know the materials you'll be working with. This includes those for bonding. Further into the construction phase, you'll be relying on this bonding — mainly on the 9410 structural adhesive — 100 per cent to hold the show together.

You'll need plastic cups, an accurate letter scale or equivalent that measures in grams and mixing stick — scrap wood will do (Fig. 3-1).

First thing is to weigh the measuring container and mixing stick. Write these numbers down for ready reference. Now, it's a matter of determining the bonding process needed. There are the 9410 structural adhesive, "5-minute" Kwik-Set epoxy and the clear resin.

The 9410 structural adhesive comes in containers marked "A" and "B" — "A" is the compound itself with "B" being the catalyst. The 5-minute epoxy also comes in containers "A" and "B," and are the compound and catalyst, respectively. Use it in a ration of 1:1. The clear resin is 5240, with a 540A catalyst. Mix it in a whatever-it-takes ratio, depending on how fast a set you want or need. The 9410 and 5-minute epoxy are used where indicated.

Using the 9410 structural adhesive, as an example: carefully weigh out 11.5 grams, not including the stick or cup. Opening the base compound, you'll notice it's extraordinarily cohesive and rather difficult to remove from the can. Scrape out a glob of this base compound to the above weight.

Figure the exact difference between your initial stick and cup measurement and the new total, which will give you the compound

Fig. 3-1. Scales and bonding materials.

weight only. Once this is known, you can figure the exact weight of catalyst needed.

Once you've got the 11.5 grams of compound, not including the stick or cup, multiply by 22.4 — 22.4 times 11.5 grams equals 257.6 grams. Now divide this figure by 100, which gives you the amount of catalyst needed. In this example, you'll need 2.58 grams of catalyst.

Or, if you prefer, this procedure can be done by volume, using a 5:1 mix. This means that for every five parts of 9410 compound, you mix one part catalyst. With a room temperature of 75 degrees F, mix the adhesive until it has a brownish-red look after about three minutes of mixing.

Naturally, the above are only recommendations of the Mitchell Wing manufacturers, as used by their personnel. Other compounds and mixtures are in use throughout the industry; however, they're not necessarily recommended. Again, check with knowledgeable personnel before trying anything out of the tried and proven method.

4

Building the Mitchell Wing B-10

To start with, look through the drawings you've purchased, making sure they're all there — assembly manual, material lists and photos. Select the #0 rib blueprint, also called the "center" rib. Lay this rib plan onto your workbench. Cover the plan with a clear piece of plastic overlay, remove wrinkles, and staple the perimeter using a very light staple gun, or an ordinary office stapler will suffice.

Tools Required

The tools needed to assemble these superb fun-flying machines are those everyday tools normally associated with the average garage workshop:

- ☐ Bandsaw or jigsaw.
- ☐ Belt sander or disc sander.
- ☐ Screwdriver
- ☐ Pliers.
- ☐ Welder.
- ☐ Hammer.
- ☐ 10″ Table saw.
- ☐ Hot wire.
- ☐ Small clamps.
- ☐ Chisel.
- ☐ 6″ or 8″ Wood plane.
- ☐ Miter saw or box saw.
- ☐ Stapler.
- ☐ Hand-operated pop rivet gun.

☐ Jointer or planer.
☐ Paint/glue brush.

If you're going to build from a kit, rather than a set of plans, you can forget about the need for the 10″ table saw and jointer/planer, as all parts are preformed and precut.

Your next consideration is an area from which your flying funster is to spring. This presents no problem, either. Subassemblies can be put together in your garage, basement, attic, living room or bedroom — anyplace that measures 10′ × 12′ is sufficient.

Once you've selected an appropriate area, you're going to need something to work on. Take a piece of plywood — a 4′×8′ sheet — and cut it in half lengthwise, making it 2′×8′. This gives you a thick, heavy, working platform from which to form your aerial joy. Prop it on both ends with saw- or work-horses, and you're ready to "do-it to-it."

Spar

Choose a length of ¼″×¼″ stock. When making this selection, hold the stick by one end, and wave it up and down, gently. This is done to detect any flaws in the wood you might have missed earlier. If it breaks, take a longer piece and repeat the procedure until you're satisfied with a good piece. Follow this routine with every selection. Remember, go easy. You're not trying to break it, only to find flaws that aren't visible — knots, pitch pockets, etc. Also, don't discard anything smaller than six inches, as you'll soon find these pieces can be used in many other ways.

Drive finishing nails along the inside contour lines of the rib drawing. Driven to proper depth, nailheads should be sticking up about ½″. Now place the length of chosen spar wood against these nails, over the actual print. Again drive finishing nails, but this time along the outside, thus holding the stick in position. Drive the finishing nails in place so as not to squeeze the wood too tight — just enough to hold in position. This then becomes a working jig (Fig. 4-1).

Place the upper and lower rib caps between these errect nails. Make sure you've nailed the jig properly every few inches. With a pencil, mark the wood where the drawing indicates "spar begins" and again where it tapers off at the trailing edge (chordline cut-off).

Gently remove both upper and lower spar capstrips. Using a razor or band saw, cut off wood at penciled marks. Cut all work slightly oversized, then sand to fit.

Replace the cut and sanded wood in the jig. At this point, you should be able to barely see the blue outline of the rib drawing to

Fig. 4-1. Rib jig— components in place for gluing, gusseting, and stapling in position.

the outside of your work. The triangular marks on the drawing are for the intersection of the vertical and diagonal truss members and capstrip crossbracing.

Begin trussing at the line indicated "rear face of spar." Fit a vertical truss to the right, so the line is just visible. Pencil mark (witness mark) the correct truss length, remove and cut. Remember, cut oversize and sand to fit. When sanding trusswork, don't round the edges. Replace the cut and sanded truss in position, then place a weight on each member to hold in place.

Now you have the upper and lower spar rib capstrips and vertical truss. Next comes the inner crossmembers. The first diagonal truss in positioned from the bottom to the top. When making corner joints, the diagonal truss is tapered so one face glues to the vertical member, while the other end is glued to the capstrip. Other diagonals are also tapered on the ends, but they mate to the capstrips only. Following the plans, once you've cut and sanded all truss members. Position them, making sure that you haven't overlooked any intersect points. Witness mark all members.

There are also diamonds outlined on your drawings. These indicate gusset plates. With a heavy set of scissors, cut, fit and sand gusset plates and trailing edge gusset plates — the latter are slightly different. Witness mark each gusset. Witness mark everything you turn out. This way there is less confusion when you get on into the project, and there are parts and pieces all over the place.

Make up a 5:1 epoxy mix, using the 9410 structural adhesive — 10cc laminating resin to 2cc hardner — and mix thoroughly (see bonding chapter). With a brush, dab the epoxy onto the tips of capstrips at the trailing edge, then dab the epoxy onto both ends of the vertical member, and replace the weights. Continue the epoxying process with the diagonal trusses, using the witness marks as your guide.

Upon completion, glue the gusset plates in position and carefully staple them to the ribs and trusses. Don't staple directly in line with wood grain, but offset the staple as much as possible to avoid wood splitting (Figs. 4-2 and 4-3).

Once you've epoxied and stapled one side of the spar rib, remove it from the jig, and turn it over to repeat the process. If there's any separation between the members, squeeze them together while stapling. You may want to practice gluing and stapling a few odds-and-ends together before actually trying your first rib.

The most important ribs are #0, #1, #6, #8 and #11. You need a pair of each rib, except for one #0 and four each of the #6 ribs. The #0 and #1 ribs are made of ⅜″×⅜″ stock, with 2″×2″ gusset plates and trailing edge gusset plates. On the four #6 ribs, a hinge block is installed three-quarters of the way back. The 1 ¼″ plywood overlays go on both sides and are angled according to plans.

On the outboard ribs — #6, #8 & #11 — a gusset block is installed between the capstrips for the stabilator hinges (Fig. 4-4). Fit carefully, and glue into position with gusset plates on both sides. All four #6 ribs have their first vertical member doubled. Apply glue to the mating surfaces. The rear spar passes through rib #0 and both #1 ribs. Two vertical members are placed to the outside of the triangular marks, with overlays on both sides.

It'll take about three hours to complete the first rib. The second rib will be finished in about half that time. And, believe me, by the time you build the 25th rib, you'll do it in 20 minutes or less. To build the complete set of 25 ribs will take approximately 12 man-hours.

Make sure you mark each rib to indicate its top and numerical sequence. Set all the ribs off to one side, and let them dry for a couple of days. In the meantime, you can be getting on with the next step.

Once the epoxy has set up, you'll want to go back and remove all the staples so carefully placed. It's going to take almost as long to remove them as it did to build the set of ribs.

The reason for this staple removal is that they are made of steel, and thus expand and contract at a different rate than wood. If

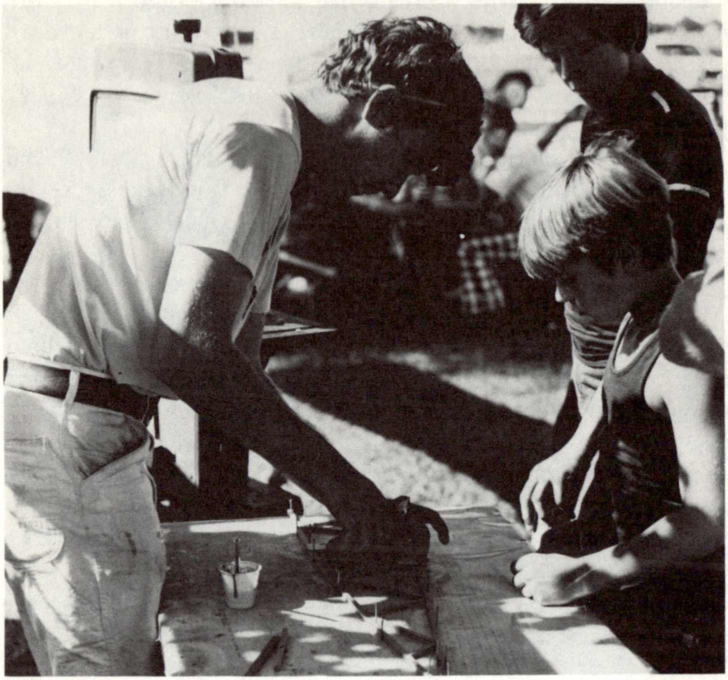

Fig. 4-2. Gusseting and stapling.

you've ever noticed a piece of wood with nails in it, especially after being exposed to the weather for any length of time, you'll also have noticed little cracks radiating out from where the nails were driven in. Those cracks were caused by the different expansion and contraction rates.

Fig. 4-3. Notice finishing nails holding rib in place.

You'll end up with about a pound of metal staples. Looking at it a second way, it's a pound of nonessential weight, and that's a pound that can be better used — 1/6 gallon of fuel, or an extra 10 minutes of flying time.

Metal Parts

While waiting for the epoxied spar ribs to cure is a good time to cut and assemble the B-10's metal components.

Plans come with templates already made for the mostly aluminum metal used with the Mitchell Wing B-10. You can simply put the templates over the metal and draw them out. Be sure to arrange templates to minimize material wastage. However, it's recommended that these paper templates be rubber cemented to scrap 1mm birch plywood, and cut out for use as sturdier templates. Paste the pages directly from the direction booklet onto the pieces of birch, then cut them out for use against metal.

Use a band- or jigsaw for cutting the metal. Timewise, you're looking at two or three hours. Make sure there are no sharp or jagged edges on the finished metal parts. Burred edges allow cracks to start. Sand and smooth up all edges and assemble according to directions. Then, as with the ribs, temporarily store them out of the way for later use.

Wingtip Rudders

Rudders on the B-10 are located on the wingtips; hence they're often referred to as *wingtip rudders*. Their location is both practical and efficient, especially since dealing with a flying wing with no aft fuselage extension or tail assembly.

Take out the full-sized rudder drawings. You'll find that there are six ribs to be cut — three for each rudder or two each of the

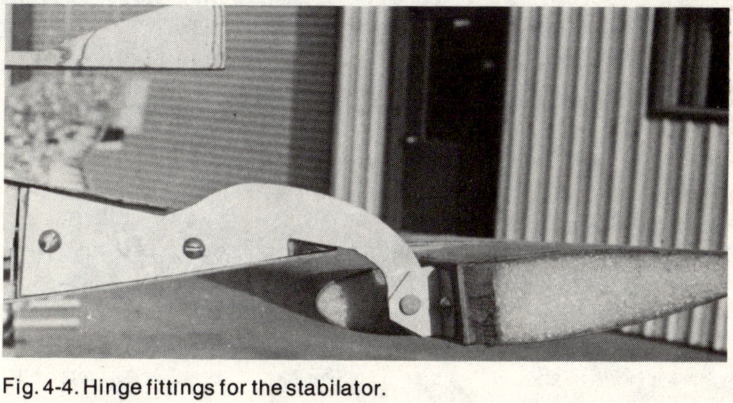

Fig. 4-4. Hinge fittings for the stabilator.

three sizes. The ribs are of ¼″ spruce or fir and are solid, thus requiring no trusses or upper and lower rib caps.

You'll need two leading and two trailing edge supports—one of each for left and right rudders. The leading edge vertical rib support is of ⅝″ square material, and the trailing edge's is of ¼″ × ½″. These vertical supports should be 22 ¾″ in length.

It'll take about six hours to build both rudders, using the epoxying and stapling methods described earlier with the spar ribs. Simply cut out the ribs full-size from the drawings, making two of each. Drill ⅞″ diameter rudder tube holes in rudder ribs at indicated points.

There are two different gusset configurations that will bond the trailing and leading edge vertical rib supports, 24 total — eight of one type and four of the other, per steerage unit.

Assembly is as simple as laying out the ribs, leading and trailing edge rib supports and gussets. When bonding the ribs to vertical supports, the #2 rib will be 11″ from both the #1 and #3 ribs. Epoxy gussets in place and staple.

When completed, you'll have two wingtip rudder units. Make sure that the flat surface of the rudder faces inside, toward the opposite wingtip rudder.

Set these assembled rudders off to one side for fabric covering, doping and attachment to metal parts. Now you've got the ribs, metal parts and rudders. You're ready to tackle the stabilators.

Stabilators

Page #4 of the plans deals with stabilators. You'll find there are two — left and right. With the exception of the end view, the plans are not to scale, as their combined length runs close to seven feet.

After cutting all six covering skins, you want to mold them into their approximate final shape. To do this, first wet all six with a 2″ swath down their center, where they're going to bend over the stabilators' leading edge. It may take two or three such soakings to get them just right.

Once they're wet enough to be pliable, try to work them into the leading edge shape. Flex them with your hands. Don't rush it or overply the pressure, as these skins will crack rather easily.

When you've got them close to the desired shape, clamp or band them until dry. One way of clamping them is to use two parallel boards about 2″ apart. Put the skins' loose ends between these two boards. Then gradually adjust the boards from 2″ to ¼″. Leave in this makeshift vice until dry.

Another way of achieving this "U" shape is to cut contour blocks. After wetting/soaking the skins, as described above, enclose them with the contour blocks, and let dry.

While the skins are drying to shape, cut the foam ribs with the use of the drawing's end view on the same page. You'll cut 38 ribs in all — 19 for each side — out of ¾″, two-pound foam. From these 38 complete ribs, you'll eventually end up with 34 nose ribs — 17 for each side — and 38 trailing edge ribs — 19 for each side. The nose and trailing edge ribs are separated by a ¼″ spar running the length of the stabilator.

Excluding the ¼″ spar, cut the entire rib 8 ¼″ in length. The finished rib size, when sanded, will measure out 8 ⅛″. At six inches from the trailing edge, cut the rib vertically. This will give you a 6″ trailing edge rib and a 2 ⅛″ leading edge rib.

Beginning on the aft side, and inboard end of the spar, place the two 4 ½″ reinforcing blocks so that they're centered at 2 ¼″ and 94 ¾″ respectively. Place the 4″ reinforcing blocks 39″ from the inboard end of the spar. Bond these reinforcing blocks in place with 9410 adhesive. After they dry, sand off any excess, being careful not to round off the spar edges.

The nose ribs can be glued in place with either a laminating resin or 5-minute epoxy (see bonding section). On the forward edge of the spar, centered from the inboard end, place the foam nose ribs at 3 ¼″, 9″, 14 ⅝″, 20 ⅜″, 26″, 31 ⅝″, 37 ¼″, 40 ¼″, 46 ¼″, 52 ⅜″, 58 ⅜″, 64 ⅜″, 70 ½″, 76 ½″, 82 ½″, 88 ½″ and 94 ⅝″.

Once the nose ribs have dried in place, they should be uniform. If not, make a nose rib template out of 1mm ply. With a Surform or sandpaper, shape to desired size, taking care not to overshape.

Now back to the preshaped skins. They're to be glued in place with the same laminating resin as used above. They'll now be cut to final sizes of 6″ × 35″, 6″ × 48″ and 6″ × 12″. The 35″ inboard skin is the first to be glued in place. Next, the 48″ section is spliced onto the rib 48″ from the inboard nose rib starting point. The 12″ skin overlaps the end of the 35″ section and the beginning of the 48″ portion.

There are various ways to get the skin down tight on the nose ribs. One way is to use the female contour blocks already made from cutting the male contour blocks used for "U" shaping the skins. Simply press them down over the skin and staple or clamp skins to the spar.

There's a need to coat the entire inside surface of the skin, face, sides of the spar and each rib with resin. Make sure a good bond is maintained with the spar.

The trailing edge ribs are now glued in place behind each nose rib (Fig. 4-5). It's necessary to trim the excess plywood flush with the spar before actually bonding these ribs. There are two extra trailing edge ribs added on the very ends of each spar for reinforcement.

The 1″ × 8″ rib caps are now bonded top and bottom (Fig. 4-6). There are 52 such rib caps. Lap the leading edge about ½″. Clamp or staple in place. Behind each fitting, the rib caps span two ribs. Put these on after the stabilator is mated to the wing, to allow access to the nuts that hold the fittings.

The two ½″ × ¼″ × 96″ stabilator trailing edges can be put on as you bond the rib caps or may be left until last. However; these edges must be kept straight. For this reason, it's recommended that you install these edges as the rib caps are put on. Clamping the rib caps to the trailing edge is preferred. Once installed, trim all excess plywood, and mate to the wing.

To finish up with the stabilator phase, carefully drill the spar, allowing room for movement between the fittings and the bracket. You'll come back to the stabilators when attaching and bonding metal fittings. For now, though, set them aside with the main ribs, metal parts and rudders.

Spar Assembly

The "D" Tube Spar Assembly is one of your most detailed and integral parts of the Mitchell Wing B-10. It's the basis from which the other structures are attached. First thing on the agenda here is the spar assemblies. You're going to need eight spar caps — four for the right and four for the left wing.

For simplicity, take the right side first. Here you'll need two inner spar caps and two outer spar caps. The inner spar caps are made ¾″ × 1 ½″ × 205″, tapering to ¾″ × ½″ × 105″. The taper begins at rib #1, or 21 ¼″ from the scarfed (beveled) end, running to the #6 rib. The outer spar caps are made ¾″ × ½″ × 96″, tapering to ¾″ × ¼″ × 96″. The taper begins at the #6 rib, running to rib #11.

Take the spar caps, both inboard and outboard, and, one by one, sight down each of them to identify which side is the tapered edge. Eyeballing down each spar cap, you'll see that one side is straight and one side is tapered. The tapered side goes to the inside

— that is to say, the flat side is the side the skin will eventually be attached to, and the tapered side will be inside the wing itself. Mark each spar cap accordingly for reference.

At the outboard ends of the inboard spar caps, and at the inboard ends of the outboard spar caps, reinforcement blocks are laminated on. Following blueprints, cut these reinforcements from blocks of ¾″ × ¼″ × 36″ stock — one block each for inner and outer spar. This is where the wing attach fitting will ultimately be bonded.

The beveled end of the blocks are away from the spar end. They're laminated to the inside (wavy side) of the spar caps with 9410 structural adhesive. At this point, check to make sure that the spar caps at this end are ½″ thick.

Spread a sufficient amount of the adhesive mix on the reinforcement blocks; then, with two C clamps, bond to the inside of the spar cap, mating the squared edge flush with the end of the spar cap. Follow this procedure with the four right wing spar caps, making sure that some of the adhesive squeezes out from the joint. *Note:* Don't overclamp.

Using a 1″ plywood sheet, draw a straight line down one 8′ side, about 2″ from its edge. Now nail small wood blocks on the outside of the line every few inches. These will hold the top spar cap straight during assembly, so work carefully. These blocks can't exceed ⅝″ in thickness, or they'll interfere with the spar web bonding.

On one end of the straight line, draw a perpendicular line about 1″ from the edge of the plywood. Draw this line 10 ½″ long. This line will indicate the inboard face of rib #1, and will also

Fig. 4-5. Aligning stabilator foam ribs.

Fig. 4-6. Stabilator reinforcement gussets installed.

determine if it's a right or left spar. Now draw perpendicular lines every 4 ⅝" from the line indicating rib #1. There will be 20 spaces between the #1 and #6 ribs. See bottom right of page #7 of the blueprints.

Take the inboard spar cap and mark a line 13" outboard from the scarfed end. Then, with the tapered edge inside, line up this 13" line on the spar cap with the line representing the inboard face of rib #1.

Block to hold in position — Be sure not to interfere with the foam verticals which will go on the outboard side of each of your lines. For now, leave the very end of the spar cap from rib #5A free of jig blocks.

From the intersection of the top of the spar cap and the #1 rib line, measure down 10 3/32", and make a mark. Then, from the top of the spar cap at the #5 rib, measure down 7 7/16", and make another mark. Now, with a straight edge, connect the two marks, continuing out an equal distance with the top spar cap. Block the lower spar cap in position, with the taper facing the top spar cap. The two scarfed edges will be flip-flopped.

With the blocking of the upper and lower spar caps, you have a jig with which to complete the spar for the right wing. Now cover the jig with a plastic or wax paper to insure bond release.

At the narrow (outboard) end of the spar caps, between ribs #5 and #6, place ¼" × ¾" wood verticals, as well as 1" × ¾" reinforcement blocks (inboard only). It's important to insure that the 1" × ¾" blocks are toward the bottom cap strip. Use laminating resin throughout spar construction. Block to hold in position.

Fit and glue the 1″ × 1″ foam verticals on the outboard side of the lines, working from rib #1. The vertical at rib #5AB has to be fitted between the reinforcement blocks and wood verticals. After the glue is dry, sand the foam flush with the cap strips, but don't overdo it.

Skin the spar with a #1 and #2 1mm birch plywood web by squeegeeing laminating resin, about one cup, on the entire surface of the web, as well as the foam verticals and the edges of the cap strips. Clamp, weight or staple the web, making sure you have a firm bond. Then glue and place the 2″ wide 1mm plywood over the splice.

After the spar and web have mated and dried, trim overhanging plywood and sand off excess glue. Next, glue and staple the spar end doubler at the uncovered end of the #6 rib. Let it dry, then trim excess, also.

Strip the jig so that only the initial row of blocks remain. This is preparatory to working with the outboard spar.

Redesignate the #1 vertical line used with the inner spar as #6. Block the top outboard spar cap in place. Remember, taper cut goes on the inside. From the #7 rib — fourth rib out from #6 — measure down 6 3/16″, and from rib #11 measure down 3 ½″. As done with the inboard spar cap, connect these two points by extending a line to the end rib. Now block the lower spar cap into position, with the line on the outside.

At the #6 end of the spar, place two ¾″ × ¼″ verticals between the reinforcement blocks (Fig. 4-7). Cut and fit foam verticals, splitting and tapering the vertical at rib #6AA.

Sand flush, after drying, and then skin. Make sure to cover the web with laminating resin, as well as spar cap edges and foam verticals. Let these dry, also, then trim off the excess. Place the doubler on the front (uncovered) side at the #6 end.

To check the right wing outer and inner spars, lay them on the floor, foam side up, butting each other at rib #6. At the inboard (widest) end of the spar, the right angle indicates the top of the spar, the tapered side being the bottom. Looking at the spars from the bottom, with the wide #1 end to the right, you've made a right spar. You're now looking at the front of the right wing.

To make a left wing, lay it all out from the opposite side of your straight line, and repeat the complete procedure. Remember to mark each of the spars right or left for later reference and to whether they are inboard or outboard.

Wing Attach Fittings

The "wing hinge point" is detailed in the upper left corner of sheet #7 of the blueprints. You've already cut the eight metal fittings and laid them aside. You're now ready to use them.

At the #6 end of the inboard spar, wood shims are laminated with 9410 adhesive on both sides. They should overhang the fittings by at least ¼", spar web sides only.

Nail 2" × 4" blocks on one end of the plywood sheet to elevate the inboard spar, foam spacers facing up. Center the #6 end of the spar on the plywood, and secure (Fig. 4-5). With a straight edge, draw a line extending out from the top of the spar. From a point ⅝" from the edge of the spar, draw another line six degrees up, continuing it out to the end of the plywood. This indicates the six degrees of dihedral to be set in the outboard spar.

Block up the outboard spar, with the top flush to the line and the #6 end — which is approximately 1 ¼" from the top of the inboard spar. Secure to the jig, but leave room for adjustment. Clamp the .090 fitting in place, top and bottom, and check to make sure the shims are going to allow clearance.

Put a 3/16" bolt through the fittings, and adjust the inboard fittings according to the measurements on the blueprints. Check

Fig. 4-7. Spar web with foam support braces.

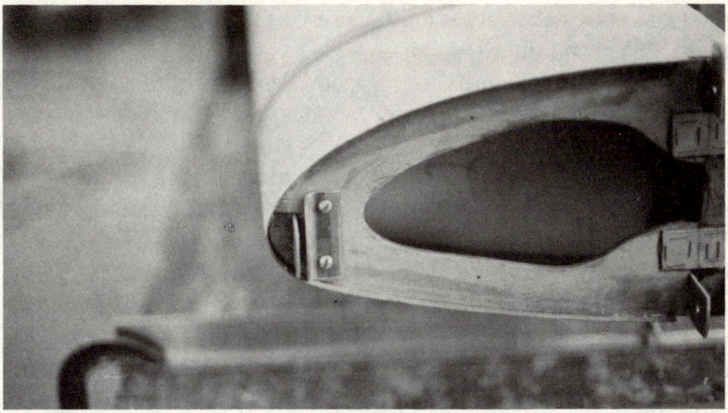

Fig. 4-8. #6 plywood nose rib and #6 rib drag attach fittings.

the shims for enough clearance — 1/64″-1/32″. Sand or laminate scrap birch plywood to fit.

The measurements of the lower fittings might vary slightly from 13/16″-1″. Again, check the angle between the two spars at the top capstrip, and mark the fittings and their locations. Drill the two 3/16″ holes in each of the four assemblies. Remove the fittings and deburr the drill holes.

Sand the inside surface of the metal to the edge of the spar and the wood. It's important not to touch the surfaces after sanding. Bond immediately after sanding.

Now carefully mix a batch of 9410; about 20 grams will do. Without removing the spars from the jig, spread a sufficient amount of adhesive on both the fitting and the spar, and position it on the top; spread adhesive on the bottom of the spar and the corresponding fitting.

Position the assembly with two 3/16″ bolts and nuts. The bolts on the inboard side are longer, to accommodate the shims. Tighten the nuts until some glue squeezes out, and the threads are in the nylon stop.

After gluing and bolting the fittings, check that you have maintained the six degrees of dihedral. Line ream the holes to ¼″, and install the hinge pin on the upper fittings.

Make the other wing by securing the inboard spar to the opposite end of the jig, and repeat the procedures just gone through.

Cut holes in the spar web between each rib station with a ¾″ or 1″ hole saw. Start from the foam side, and finish the hole from the back, using a block to reinforce the web. Trim excess plywood at the #6 rib end and along the capstrips.

Trace the spar onto the plywood sheet, and cut ¼″ inside the line so that the edges of the spar hang over. Do this to the inboard spars first.

Drill holes in the plywood and sand so the attach fittings aren't distorting the actual spar. Nail 2″ × 4″s on the edge flush with the long sides, and put nails even with the foam every 4 ⅝″ through each of them. Leave the nail heads ½″ high. These are for the surgical tubing that holds the skin in place while the glue is drying.

Bond the reinforcement blocks to the end of the leading edge with 9410 adhesive. After they dry, sand the mating surfaces of the leading edge attach fitting, and bond with 9410, according to detail drawings at the top right of sheet #7 of the blueprints.

Secure the spar to the jig, making sure the nails are even with the foam which is facing up. Take one of the four ¼″ plywood nose ribs, and determine which is the top of the spar, referring to the detail at the top of print sheet #7. Sand the rib to conform to the pattern. The wood rib goes flush with the end of the spar.

You must notch the ends of the rib to clear the attach fittings and the shims. Also, the correct leading edge must be fitted ½″ outboard, and ⅝″ inboard, according to plan.

With 9410 adhesive, glue the rib to the spar, using small 1mm plywood gussets glued and stapled to hold it to the spar. Run a thin fillet of adhesive between the spar, fittings and rib. See detail at top right of sheet #6 of plans.

Using a framing square, hold the rib 90 degrees to the spar, bracing it in place until the glue dries (Figs. 4-9 through 4-12). Check to make sure camber is on top. Label the white foam ribs from #1 (largest) through #40 (smallest) on each set of spars.

At the opposite end of the #6 rib, epoxy the appropriate foam rib to the spar and spacer, using either 5245 laminating resin, or 5-minute epoxy. Again, double check to make sure the camber is on top.

Now make a template out of wood or masonite of the #1 rib, using the patterns on sheet #5 of the blueprints. With nails or a bit of wire, secure the template to the rib. Check to see that the template matches the width of the spar and the notch in the foam also matches.

Place the leading edge in the notch, to help you align the foam ribs (Fig. 4-13). Glue the ribs to the spar and spacer, keeping the leading edge straight. Check for sequence and top of camber.

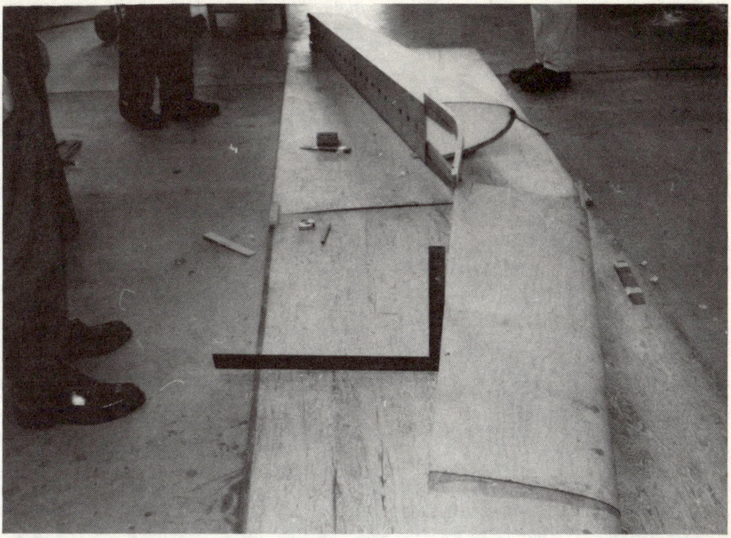

Fig. 4-9. Carpenter's square being used to assure rear face of spar is 90 degrees.

After the glue sets, epoxy them to the leading edge, taking care that they are perpendicular to the spar. It's important that the leading edge is bonded to the #6 rib with the 9410 adhesive. Make sure the leading edge and reinforcement blocks are flush with the outside edge of the #6 rib, with the fitting sandwiched between them.

Fig. 4-10. Rear face of spar must be held 90 degrees.

Fig. 4-11. Typical blocks affixed under leading edge to hold 90 degree angle.

Get a straight, uncut length of 2″ × 4″, and staple sheets of heavy grit sandpaper on the face for about 90″, with handholds on each end. On one end, staple a 1mm plywood strip as a guide to run over the #6 rib, as you sand the foam ribs to shape. Before you start to sand, round the leading edge to conform to the #6 rib contour. Sand the ribs slowly, keeping the sanding board in the same position on the end ribs (Fig. 4-14). Sand gently, so that the

Fig. 4-12. Once leading edge is shimmed, it must be held in place.

Fig.. 4-13. Connecting outer panels with nose rib capstrip.

ribs aren't broken loose. When finished sanding, the leading edge spar should conform to the indicated airfoil shape.

Before skinning, you want to install the control box built earlier (Figs. 4-15 through 4-17). It would be impossible to install it after skinning, without cutting holes in the ship's leading edge. Also, attach and secure all metal fittings for the tricycle landing gear cage (Fig. 4-18). See pages #2, #3, and #9 of the plans.

Check that total contact will be made with the skins, especially where wood and foam meet. There are two skins per spar. Make sure you choose the proper skins.

Fig. 4-14. Sanding foam nose ribs to shape.

Fig. 4-15. Control box reinforcement and nose rib capstrip in place.

Skins are cut oversize to permit adjustment. Starting at the wood rib #6, lay the first skin over the foam and strap with surgical tubing. After adjustments, there should be overhang on both the top and the bottom of the spar. Also, there should be overhang on the end of the spar. The other edge of the skin should be on a rib. Mark for the upper fittings, and trim.

With the first skin still on the spar, lay the second skin over the foam and adjust (Fig. 4-19 through 4-23). The skins need to

Fig. 4-16. Typical mixer box installation.

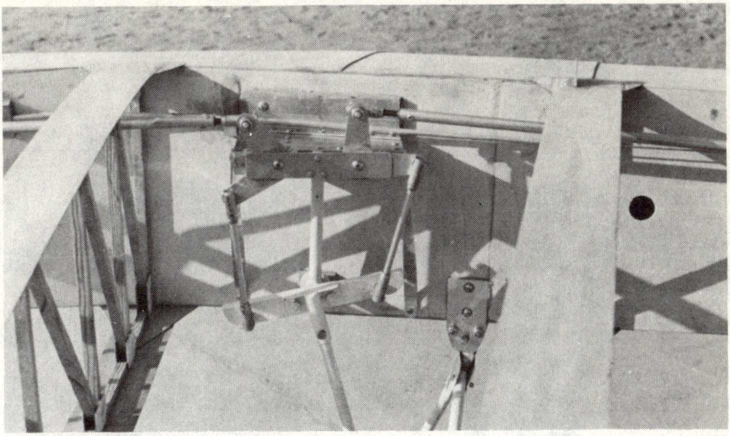

Fig. 4-17. Mixer box assembly.

overlap by at least one inch. It might be helpful to tack two nails through each skin into the leading edge to align them after the glue is applied.

Surgical tubing must be strapped over the skin at every rib station. The tubing must be tight enough to insure a good bond between the skin and the foam. If your area is particularly dry, moisten the skin in place and let it dry before gluing. It's important that every item be ready to go before any glue is mixed — about ½ cup of the laminating resin per skin will do.

Paint each rib edge and the entire spar cap up to the overlap with the mixture. Spread an even coat over the skin, and then lay it

Fig. 4-18. Installing cage attach fittings.

Fig. 4-19. Inner panels ready to receive skins.

in position over the foam, using the nails as guides. Strap in place with tubing at each rib. Staple the skin to the spar every ½″, except for the 1″ overlap.

If you can't get the skin to shape properly, put some staples in the #6 rib. Repeat the procedure for the next skin, making sure enough glue is applied to the 1″ overlap.

Fig. 4-20. Applying skin to nose ribs with use of surgical tubing for tight bond between skin and ribs.

After you've finished skinning, and the glue has dried, take a coin and tap around the wood where it has made contact with the foam ribs. This is done to insure a solid contact bond has been made. If it hasn't, you can readily tell from the sound. It'll sound much like a drum, instead of a hard knock, if it's loose.

If you discover an area where there's no contact, secure a large syringe with a long nozzle from a veterinarian supply store.

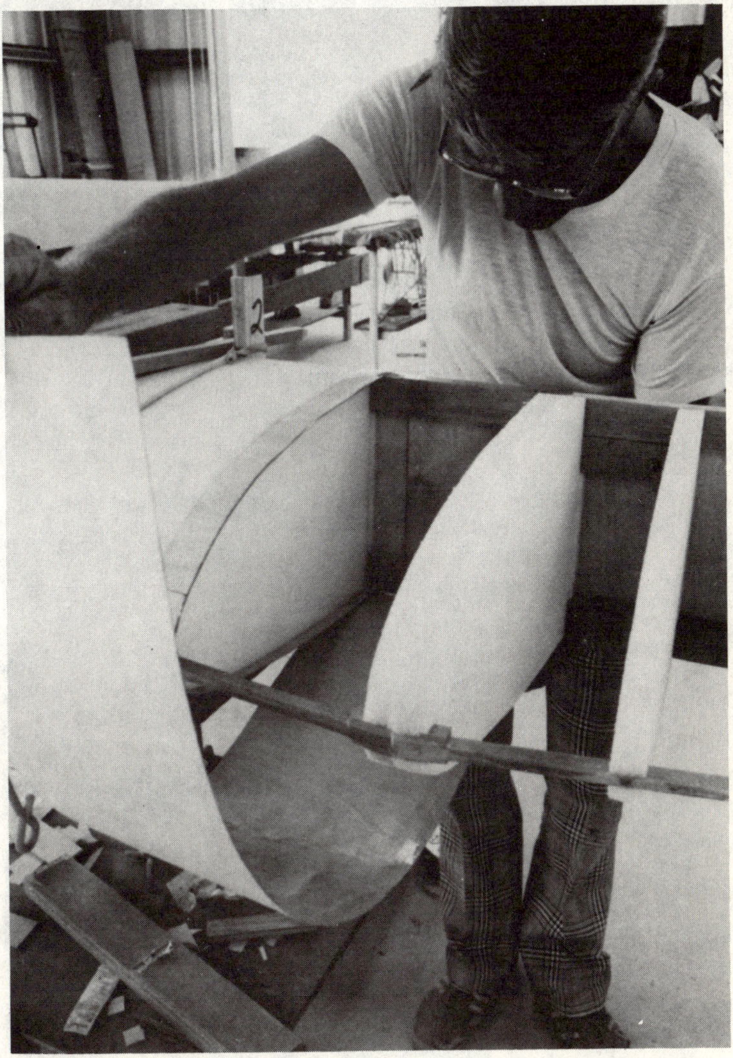

Fig. 4-21. Forming skin to nose rib shape—notice skin overlap.

Fig. 4-22. Fitting D-tube skins.

Turn the leading edge over. Load the syringe with glue, and go through the lighting holes in the spar web. Lay a coat of epoxy in there between the wood and foam. Let it dry, then tap check again.

Now you've got your inner and outer spar and leading edges built. If you assemble the components, you've got the full airplane, without any of the main ribs attached yet, of course. You can fold and unfold the wings. You're ready now to install the previously built ribs.

At this point, you can sand the leading edges and get them nice and smooth (Fig. 4-24). Also, it's here that you take care of any

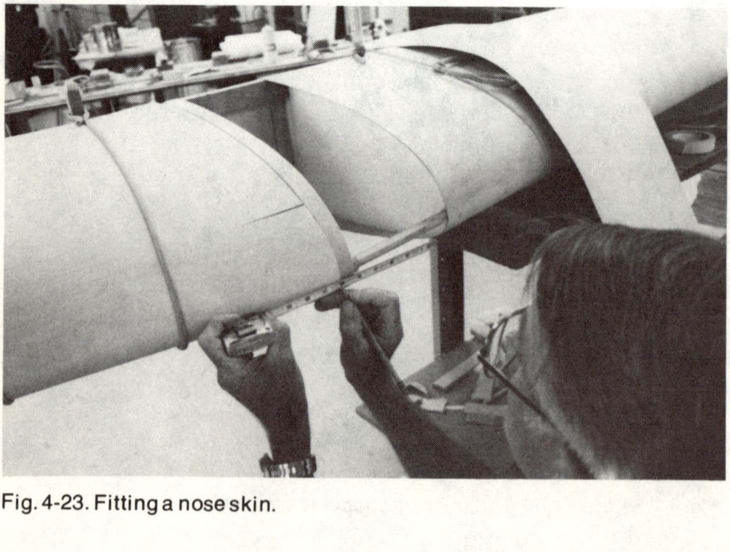

Fig. 4-23. Fitting a nose skin.

mishaps by patching or epoxying on it. Get the sections just about ready for what you could say is "paint." It's also during this phase that you want to connect the inner spars with a laminated "bow," and install the center section nose ribs.

Laminated Bow

Refer to pages #6 and 11 of blueprints for details. Lay two plywood sheets on the floor, their ends butted together and tacked secured so they won't separate. The butted ends form a vertical, central reference line.

About one-quarter the way up from the bottom of these plywood sheets, along the reference line, make a mark. From this mark, measure up 16″, and draw a semicircle using the 16″ as a radius. Also, where this reference line intersects the semicircle, measure left and right for 18 ½″ around the semicircle, and make marks.

Take your two inboard sections (right & left), turn them upside down, and place them lengthwise on the plywood sheet in such a manner that the left inboard rib #0B is touching the left semicircle mark, and the right inboard #0A rib is touching the right semicircle mark (Fig. 4-25). Place them in such a fashion that the inboard spars' beveled (scarfed) extension runs along the inside of this semicircle.

Fig. 4-24. Sanding leading edge skins of D-tube.

Fig. 4-25. Establishing exact sequence and distance between outer panels.

If the above is done correctly, the inboard spars will be swept back the required 12 degrees, with a total of about 37″ separating the two #0B ribs (Figs. 4-26 through 4-29). The open area between will eventually house the "O"/center rib and the two "OA" ribs.

The laminated bow that connects the two inboard spars is nothing more than three ¼″ pieces laminated together in such a way as to form the open area, 16″ radius semicircle. You'll need

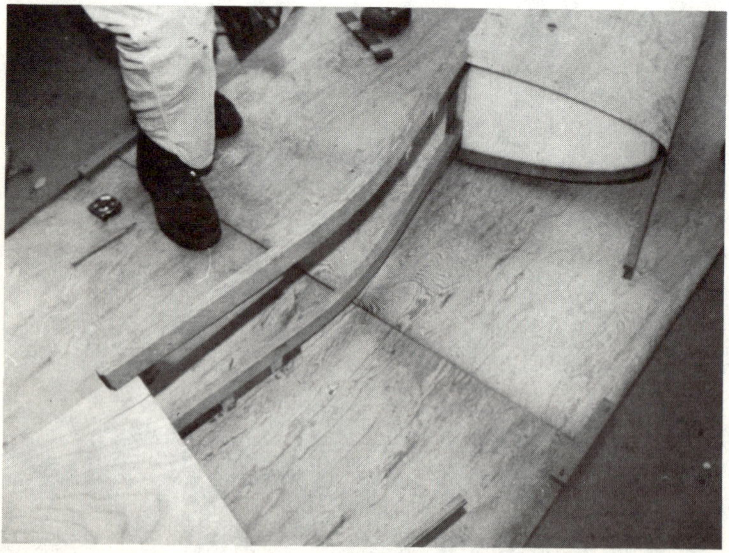

Fig. 4-26. Fitting 12-degree sweepback laminated bow.

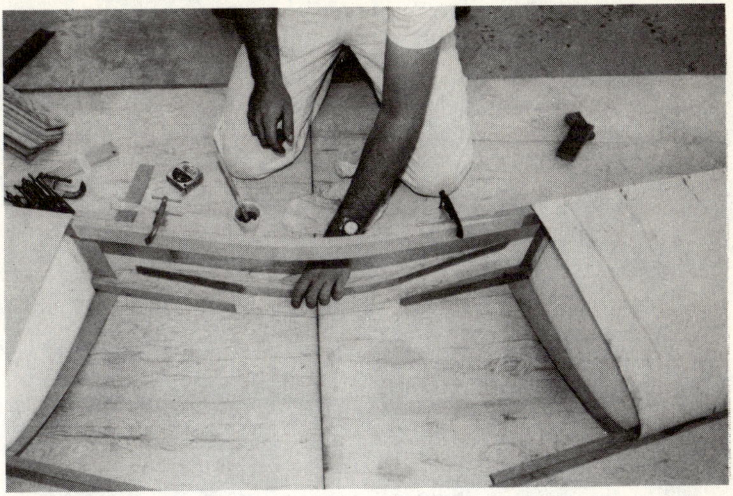

Fig. 4-27. Fitting 12-degree sweepback laminated bow.

two of these ¾″ laminations — one for the top, and one for the bottom. Laminate with 9410 adhesive.

Once the laminations have dried, mark, cut and scarf these caps to mate with the leading edge cap stubs. The lower cap strip, the one farthest from the floor, because the leading edge is upside

Fig. 4-28. Fitting 12-degree sweepback laminated bow.

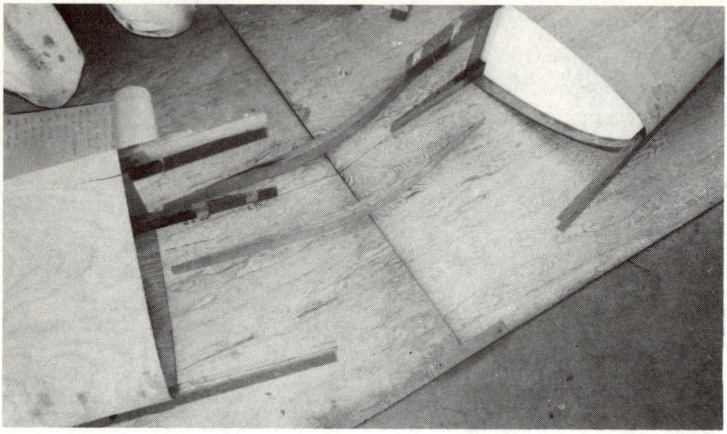

Fig. 4-29. Fitting 12-degree sweepback laminated bow.

down, has dihedral, and the laminated cap is cut so that it has dihedral, also. Make sure you mate it correctly with the stubs before actually cutting.

Now glue the scarf joints, using 9410 structural adhesive. Use six clamps per joint, with clamping blocks under the clamps for glue pressure (Fig. 4-30). Apply adhesive to both surfaces, and allow to dry.

Cut a drag strip off 9″ inboard of the inboard face of rib #1. Work carefully, as not to break the strip at the rib. Cut, fit and glue the drag strip, joining the two leading edge assemblies.

After all glue has set and the clamps removed, fit a spar web to the back face of the spar. Butt it up to the existing web to within

Fig. 4-30. Laminated bows and nose capstrip fully clamped and drying.

1/16″. Glue the web to spar caps. You may want to staple. However, make sure you start at the center line of the spar, and work outboard. Use 9410 for bonding. Once glue is set, match the two web butt-splice plates.

You're now ready to install the ribs. Place the assembly on two sawhorses with the webs resting on the sawhorses. Due to the sweep back, the assembly is very unstable, and tends to flip one way or the other. Be sure you don't damage the outboard fitting when moving the leading edge. Place the sawhorses in such a manner as the center area is at a convenient height.

Adjust the horses or shim under the spar so that the two sticks clamped to the wing attach fittings are aligned. This is very important, and they must be kept parallel until the last piece of skin is glued in place.

Now, bond in place the foam strips between the caps — one at center line; one 9″ inboard of rib #1; and one in between each of those. Use 9410 adhesive. Glue the block for the cage fitting, making sure you glue to lower cap, and not the top one.

Install the #0/center rib and both #0A ribs with 9410 structural adhesive. Fit, clean, bond and bolt the cage fitting to lower cap. Fair ribs and leading edge drag strip.

Glue on the right and left outboard skins first. Overlap the skin by at least 1″. Make sure you get a good bond by using skinning methods used earlier.

Fair ribs 0A (right), "O" (center), and 0A (left). The center section skin will overlap the right and left outboard skin by at least 1 ½″ at the spar, and will get less, as you fair towards the leading edge. When fairing, you'll have to start scarfing the outboard skin as you move towards the leading edge. At the leading edge, you should have at least ¼″ minimum bonding area. Allow to set until glue is dry.

Trailing Ribs

You're ready to install the trailing ribs now. The web area should be vertical. Clamp it to the sawhorses with bungee chords or surgical tubing so that the leading edge is held in place and not knocking about.

Measure off every 18 ½″ from the #6 rib in, according to the drawings. Take the #0/center rib, and the #6/outboard rib, and put them in place, using gussets. They're not glued yet, but simply hanging in place.

Next, using a piece of scrap wood and a clamp, support the trailing edge, with the stick running to the workshop floor. Clamp

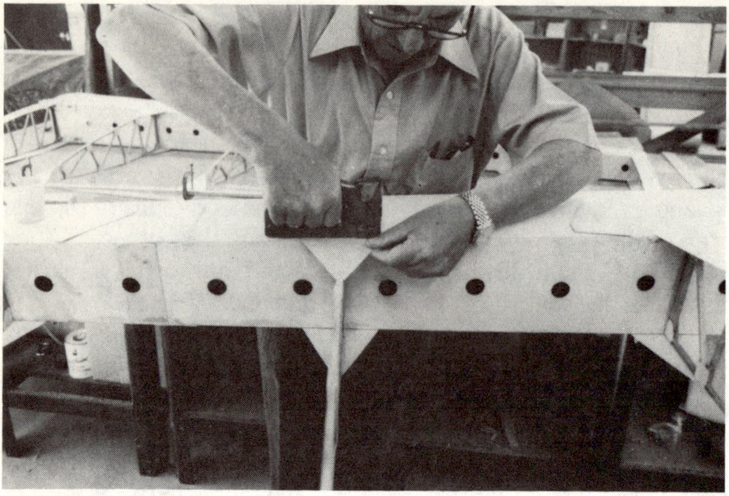

Fig. 4-31. Installing gussets for rib attach.

that trailing edge so that it's aligned 90 degrees to the spar, and that the face of the ribs are flush against the spar web, and it's supported. You might take a couple of scraps and run 45 degrees out from the rib, and staple that in place, also — so that the #0 rib is locked in place with gussets where you want it.

Go out to the #6 rib, and do the same thing (Figs. 4-31 through 4-33). Make sure that it's 90 degrees, and that it's in place, and

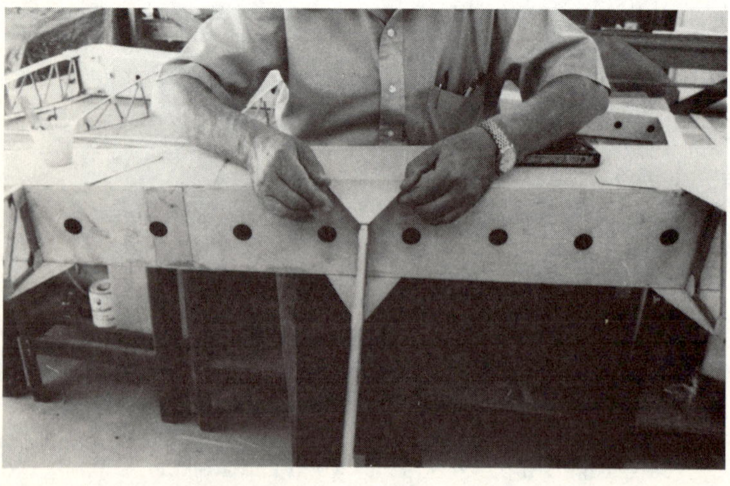

Fig. 4-32. Fitting gussets for rib attach.

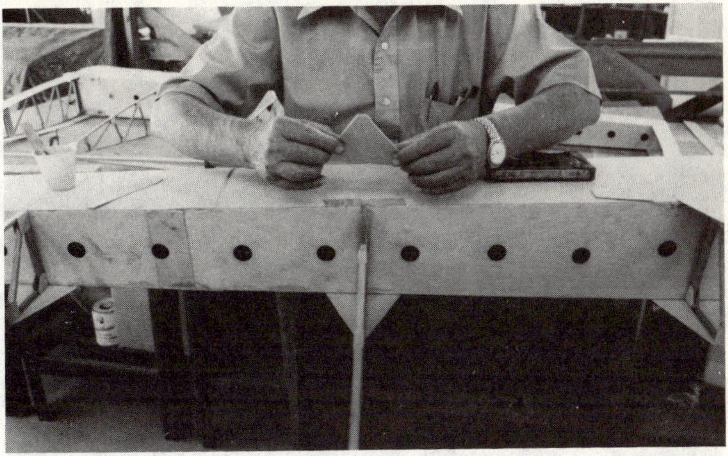

Fig. 4-33. Gluing gusset area prior to gusset installation.

everything is like you want it — per specifications — like you're going to end up with. Support it also with a scrap stick to the garage floor. It's all being held in place by staples and those two sticks.

Epoxy the leading edge of the #0 and #6 ribs against the web and the spar (Figs. 4-34 and 4-35). Now you've got the #0 rib and the two #6 ribs outboard. The three ribs are installed, and they're all lined up with the spar and epoxied in place.

Install the additional ribs every 18 ½". You're going to find that these additional ribs don't fit perfectly. Remember, you used

Fig. 4-34. Installing #6 rib. Note: center of chordline temporarily identified with a ¼ × ½ horizontal wood strip.

Fig. 4-35. Nose skin held in place with surgical tubing.

the additional sticks to support the #0 and #6 ribs where you wanted them. Well, now you're going to take a string or thread, and stretch it between the three. You'll find, when installing these ribs, that the trailing edge doesn't quite line up with the string — showing where the trailing edge should come to be perfectly

Fig. 4-36. Trailing ribs. Notice truss members and gussets.

straight. Where the foam rib attaches to the web, sand an angle on the rib to bring it down, or up to the exact level of the leading edge of the rib, where it attaches to the spar. You have a sander, so you can taper it to lower or raise the trailing edge.

Fig. 4-37. Rudder cables routed through housing and push/pull rods for control system installed.

Fig. 4-38. Snaps connecting rudder cables.

When you've got those fitted and lined up with the string, you have another problem. You're going to find that some of them are shorter or longer than others.

To solve this dilemma, you can add to some, and take off of some, or you can simply take off of all of them. Install the trailing edge, so that you end up with what can be termed as straight.

Now you've got the ribs installed. You've got nice straight trailing edges, all held in place with gussets (Fig. 4-36). Go ahead and do the outboard panels — install those ribs the same way you did the inboard.

Fig. 4-39. Spring-loaded wingtip rudder mount.

You'll need a bigger working area now, so open the garage door. Install the outboard panels, making sure all the ribs are in alignment.

If you build the panels separately, you may find that the #6 rib on the outboard panel to be ¼" higher or lower than the other one. It's better if you put them all together, and continue that string line on out, so that everything is lined up. It's here that you go back and pull the staples in all those gussets you've put on.

Now you've still got to put on the rudder brackets for the wingtip rudder controls (Figs. 4-37 through 4-39). You're going to install everything here, including the push/pull rods (Fig. 4-40) to run the control system; bellcranks at the outboard #6 rib; cables to control the rudders — outboard drag rudders; and, you've already installed the control mixer box.

At this point, you want to install the outer wing panels, and also install the stabilators. Install the rudders and rudder block assemblies. The rudders are in line of flight.

Remember, these wings are swept back 12 degrees. That means that the wingtip rudders need to be offset 12 degrees, also. Many builders insure the 12 degree alignment by striving for 12 ½ degrees. This gives the rudders a bit of toe-in, and, to a certain extent, this toe-in increases the stability of the Wing.

One of your final steps in the construction phase is rigging. To do this, install the control stick in the vertical position. Set the stabilators at about six degrees, while keeping the stick in neutral.

Fig. 4-40. Bellcrank fitting for push/pull rod.

Fig. 4-41. Final inspection by FAA prior to covering.

Pull the stick all the way back. The stabilators should automatically come up to about 35 degrees. Push the stick forward, and the stabilators should now go to zero degrees or a negative one degree — not below the flight line.

Controls should never go negative. You may have to add a block as a stop between the forward part of the stick and the web.

A counterweight is now installed on each control to minimize the chance of any flutter ever occuring in the control system. It's spelled out for you on page #4 of the blueprints.

From here (Fig. 4-41), it's just a matter of doping, fabricing, painting and choosing the right hang cage for your mode of fun-flying. See those sections for details.

Those Finishing Touches

5

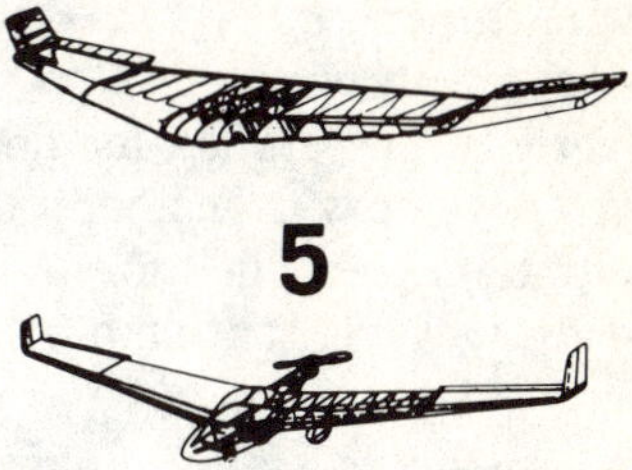

After completely assembling your Mitchell Wing, you can finally see the skeleton end of your struggles—it's beginning to look like the pictures, after all. However, before proceeding to covering it with Dacron fabric, there's another step yet—that of varnishing.

Varnishing

Use just plain old spar varnish or polyurethane, and apply it judiciously over the inside of the rear of the spar, all of the ribs and every bit of wood that's going to be exposed to weather or humidity. Put on at least two, and maybe three, coats—whatever it takes to make it look like a piece of finely finished wood.

Take extreme care that you don't varnish in the areas where the fabric is to be attached—trailing edge, top and bottom of the ribs and the aft edge of the "D" tube. Varnish and nitrate dope simply don't mix—they're not compatible, and strange things happen when they do get together.

One way to avoid this varnish/nitrate clash is by first masking off the areas described above that the fabric is going to adhere to, then varnish the whole machine so that it looks like a piece of newly finished, fine furniture inside—which it *is*.

It takes about a day for the coat of varnish to cure, so you're looking at three to four days in this process. Also there are a lot of cracks and crevices to seal with the varnish around and between the ribs and gussets.

After the varnish has dried, untape the covered/masked areas. Once the tape is off, apply nitrate dope to the uncovered or unvarnished areas that the Dacron covering is to be tacked to.

Don't limit it to simply one coat of dope, but keep adding coats, so that you have a high gloss everywhere the fabric is going to touch.

Fabric Covering

When varnishing, you masked off certain areas to prevent varnish penetration. This was done because varnish and nitrate dope simply don't mix. Now you want to cover the ship with dacron fabric in preparation for final doping and paint. However, now you're faced with another problem—the fabric will not bond to the nitrate dope applied to the masked-off areas. The Dacron fabric does, however, have thousands of very small pores (squares) that

Fig. 5-1. Beginning fabric covering.

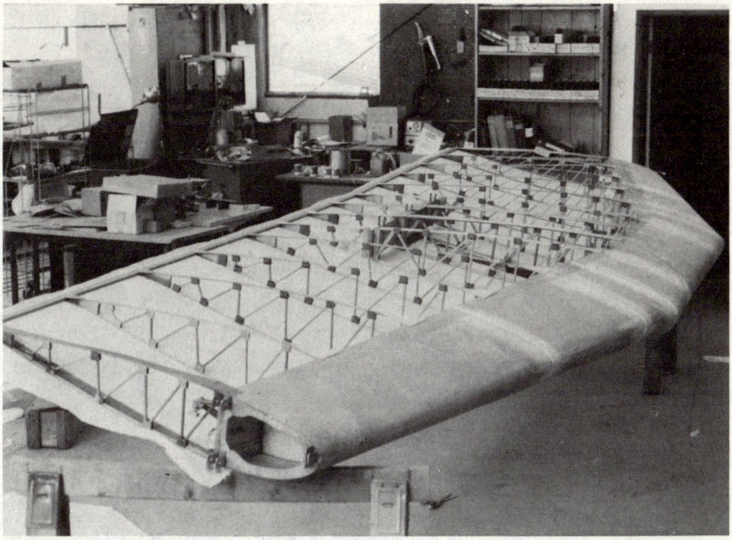

Fig. 5-2. Applying Dacron fabric.

let the dope penetrate through to the layers below. This process will become more apparent in just a minute.

It doesn't really matter where you start covering—just begin. For simplicity, start with the center section, or the #0 rib. Staple or tack the fabric to hold it, because it's going to slip and slide all over the place (Figs. 5-1 and 5-2).

Once it's tacked into position, trim off the excess at the #6 rib. It doesn't matter if you do the upper or lower surface first. With the fabric in place, you're ready to dope the perimeter—that area previously masked off, then doped.

Carefully mix thinner and nitrate dope in a 1:1 ratio. Brush the mixture on the outside rim of the fabric, on the area previously masked off from the varnish (Fig. 5-3). This thin coat of dope is going to sink right through the Dacron pores and into the nice shiny coat of dope below. In the process, the dope is going to carry this Dacron fabric with it.

If effect, it melts this fabric right down into the dope underneath. Now add more coats of dope in this contact area, until you have the fabric sandwiched in between upper and lower dope coats.

If you look at the top section of the fabric, you'll notice that its pores are full. The rib caps are all doped up like this, as well as trailing and leading edges.

Now take an ordinary house iron, or heat gun, and shrink this fabric. It's all in place now, and it's wrinkled—lots of wrinkles, sags, and there are problem areas—but don't worry about them now, as most will take care of themselves (Fig. 5-4).

When applying heat, you have to be extremely careful, because this Dacron fabric is very sensitive to it. Dacron started out as a solid ball of synthetic material. Manufacturers extruded it through machines and made it into fine woven threads. When heated enough, it has a tendency to return to that solid ball it started from.

At this point, all you want to do is heat-shrink it just enough to *barely* take out the wrinkles—it'll still have lots of sag, but all the wrinkles are gone.

Fig. 5-3. Doping fabric of wingrip stabilator/stabilizer.

Fig. 5-4. B-10 stabilator wingtip rudder covered with Dacron awaiting heat shrinking.

Add coats of dope, covering the whole wing. Brush on at least three more coats of dope. This is done to seal the pores of the fabric over the entire surface. Also, this dope will actually further shrink the fabric, but only slightly.

Let the wing sit for two or three days, and see how much actual shrinkage does occur. If you're not satisfied, again apply heat gently to bring it up to the tautness desired, but one that is still loose. *CAUTION:* If you tighten it too much, within a year, it'll do some additional shrinkage and cause the trailing edge to have a "swallow" effect. It could even pull the ribs out of shape.

Once this step is completed, you're ready to put on the color coats of paint, and apply those personal touches.

Painting

By now, perhaps, you've ground and flight tested your Mitchell and are anxious for it to be a delight to the eye. You've had plenty of time while building and flying to think about colors and paint scheme and are now ready to "do-it to-it."

The most important considerations in spray painting are the use of the proper spray gun, its spray tip and needle. Also of importance are proper air pressures and fluid viscosity for the material being applied. These can only be determined by recommendation or experimentation. Once these have been selected and adjusted, the final determining factor in the quality of a paint job is the actual application procedure.

When beginning, hold the nozzle of the spray gun 6-10 inches from the surface, depending on whether it's wood or fabric being painted. Keep it close enough to lay a good, wet coat on the surface, yet far enough away that the material doesn't run or sag (Figs. 5-5 and 5-6).

Hold the spray gun perpendicular to the surface, assuring an even spray pattern. If the gun is tilted or tipped, the pattern will be heavier on the side nearest the gun and dry and rough on the side farthest from the gun.

Move the gun parallel to the surface being sprayed. Start the stroke, then pull the trigger. Release the trigger before completing

Fig. 5-5. Painting top right outboard panel.

Fig. 5-6. Painting right bottom of outboard panel.

the stroke. This avoids paint build-up at the start and finish of the stroke. Also, keep the gun at a steady height above the material being sprayed. If the gun is arced when spraying, the surface will be sprayed in an uneven pattern, with a heavy deposit where the gun was nearest the surface and thin where the spray arced away.

Before beginning to spray the large, flat areas of the structure, do the edges and corners. This is done by spraying along the corner, which gives the thickest coat along the edge and blends out onto the flat portion.

A single layer of paint sprayed on the surface by one pass of the gun will typically be about 12 inches wide, thicker in the middle and tapered off at each end. In order to get a good, even build-up of finish, apply spray on the first pass, then, on the return pass, overlap all but about 2-3 inches of the first pass. The next pass will overlap all but 2-3 inches of the second pass, etc. Continue this overlapping process, and the results will be a nice, even paint film, with no runs or sags.

Position your Mitchell Wing in such a way that airflow moving through the area, building or spray booth will be from tail or trailing edge toward the nose or leading edge. This is done so that you can paint in the direction of the air flow, and the overspray will be ahead of you.

Start with the stabilators and wingtip rudders. In this way, you can, more or less, use these airfoils as practice for the main wing sections. First paint the ends and trailing and leading edges. In short, before starting any flat surface, paint all the difficult areas, then proceed in a systematic sequence.

Apply a minimum of three coats—primer coat, tack coat and finish coat. They're applied in different directions. Begin with the bottom or underside.

On the bottom, the primer coat is applied chordwise, from trailing to leading edge, or vice versa. Since you've already painted these edges, you'll have ample lead-in to the flat surface.

The tack coat is sprayed on spanwise, or tip to tip. This is in the opposite direction/movement as that of the primer coat.

The finish coat is put on in the same direction as that of the primer and opposite from that of the tack coat. Remember—begin spray after stroke is initiated, and end spray before stroke is completed.

After the bottom or underside has cured or dried, flip the airfoil over for the top surface. Again paint the ends, trailing and leading edges first, including all hard-to-get-to or difficult areas. From here, everything is reversed from that of the underside. In other words, the primer coat is applied not chordwise, but from end to end; the tack coat is applied not end to end, but chordwise; and the final coat/finish coat is sprayed not chordwise, but from end to end. Repeat the overall sequence with the main sections of the wing.

It's often impossible to reach completely across the wing, so spray as far as you can reach. If this is the case, keep the gun tilted back so that the overspray won't fall on that part where the paint has hardened to the point that the overspray won't blend in.

Spraying on a coat of acrylic lacquer with an excess of solvent can be used to wash out acrylic overspray. What this does is soften the film and allows the overspray to sink into the finish. Dried overspray from any material other than polyurethane can be burned down or washed out by spraying a mixture of one part retarder and two parts thinner on the surface, while the overspray and base finish are still fresh. This mixture will soften the surface enough to allow the overspray to sink in and also allow the surface to gloss. Enamel overspray doesn't usually present the problems of lacquer or dope, since it has a much slower drying rate. The overspray can sink into the finish while it's still wet.

One thing to remember: the thinner the mixture, the better the finish. For the Mitchell Wing, it's recommended a 4:1 (thinner/dope) mixture be used. This gives an extra-shiny lustre and definitely stands out.

Fabric Covering: In the Long Run

The elements that can lead to decomposition begin to attack your aircraft fabric before it's applied on your craft. These elements are like a cancer—slowly, but continually, they wear away at fabric and the finishes that protect it.

Dacron has eased the minds of many B-10 and U-2 owners who have covered their Mitchell Wing with it. It's considered by some more resistant to the environmental demons than standard Grade "A" cotton and/or Irish linen—both are organic, top-of-the-line fabrics and widely used in the antique and homebuilt fields.

Dacron is made from petrochemicals. Therefore, it doesn't respond easily to the moods of nature. This synthetic fabric has an indefinite lifetime. Indefinite in the sense it hasn't been around long enough to find out, but definitely 20 years or longer. So, it should be watched, crooned over and taken care of.

Inspect the fabric before it's put on your craft. If there's any doubt about its strength, test it.

The greatest single cause of aircraft fabric deterioration is sulfur dioxide. It's present in the atmosphere and most prevalent around industrial areas. Combining with oxygen, sunlight and moisture, sulfur dioxide becomes sulfuric acid, the same stuff that's in your battery. This type moisture or dew has often been referred to as *acid rain*.

Besides air pollution, the age-old rotting duo most ready to attack your Dacron covering consists of sunlight and heat. They act hand-in-hand, and sometimes separately, to remove the youth of the finish first, then onto affecting the fabric itself.

The finish on your carefully-built dream, therefore, is as important as the fabric underneath, because once the finish is destroyed, the natural fabric will be closely behind in deterioration (Fig. 5-7).

Oil, grease and fuel from the engine and tanks have chemicals that are harmful to the finish. With the Mitchell being a pusher, this area isn't much of a problem, but it still exists. As you tenderly wash the craft, use extra care not to use too-strong solvents or too-potent soap solutions. A good wax job will add further protection.

A light colored paint scheme is recommended over a dark one. Dark colors absorb heat more than lighter ones. And proper storage and care will add years to the skin. But regardless of your treatment of the Dacron covering, the FAA says that the fabric is airworthy until it has deteriorated 30 per cent from the originally required tensile strength.

The only true test of your covering's tensile strength is one made with a strip of undoped fabric in a laboratory. The lab inserts the fabric strip into a tester, which literally pulls it to the breaking point.

There are other tests that can be made in the field on doped fabrics, one such being the punch tester. This cone-shaped pointed object is placed against the fabric, and the unit then pressed until the fabric fails, and a hole is punched through. A reading on the punch tester scale indicates fabric strength.

You, as the owner of a fabric-covered airplane, can do much to either extend or shorten the Dacron fabric's lifespan. All it takes is a basic knowledge of what affects its longevity.

Fig. 5-7. Protecting fabric and finish the easy way.

[Editor's note: While the method mentioned must work all right, there is no need to employ the highly flammable nitrate dope. Stits Poly-Tac Aircraft Cement will give a much better bond, as will the special cement employed with the Eonnex covering system. In fact, the Eonnex covering and finishing system would seem ideal for the Mitchell Wing. It is FAA-approved. The Eonnex Dacron is available in lightweight sailplane cloth, and the Eonnex finishing system requires only three coats of water-soluble primer and finish, another weight-saving feature. The Dacron is heat-shrunk to its final tautness prior to the application of the filler/primer/finish. We would never apply a hot iron to a nitrate dope finish. Before you cover and finish your Mitchell Wing we strongly recommend that you contact Mr. Bill Lott, Eonair, Inc., 417 Watts Drive, Bakersfield, CA 93307, and/or Stits Aircraft Coatings, P.O. Box 3084, Riverside, CA 92509, for information about their products. You can also use Ceconite's Dacron system which is normally finished out with butyrate dope: Ceconite, Inc., 4677 Worth St., Los Angeles, CA 90063.]

6 Flying the Mitchell Wing B-10

The 230-15 airfoil is probably the finest wing ever developed for good lift and inherent stability. The center section of the wing is flat, with two outboard panels set at 6-degrees dihedral, and the wing is swept back 12 degrees from center. The sweep back and dihedral, along with wingtip rudders, give the aircraft exceptionally good directional stability.

The washout that is required to prevent any flying wing from tucking is accomplished by setting the stabilators at a negative setting of 2-36 degrees, with no positive setting at all. Remember, you're working with a reversed (upside down) airfoil with the stabilator. Therefore; what normally is thought of as positive is, in fact, negative.

The rudder control, the control stick manipulating the stabilators through a mixer box—also causing them to act as elevators—results in a fully controlled ultralight airplane, with inherent stability, and one that's easy to fly.

The rudders, acting separately, give excellent turn control, or they can be activated together to produce drag for an increased descent rate as spoilers, air- or speed-brakes.

The center of the wing always stalls first, while the outer wing panels are still flying. This prevents the aircraft from stalling/falling to one side when the center section does reach the burbling point and stall.

The proven structure—the 230-15 airfoil, the full three-axis controls, and, most of all, the pilot having full control, without using the body for pitch and yaw control—results in a responsive, stable aircraft, with performance second to none.

Because of the rapid development of power systems, and the technique of its specialty, Mitchell Aircraft Corporation doesn't normally supply the Mitchell Wing with an engine, but, rather, supplies information on engines and their suppliers. All engines recommended by Mitchell have been competently tested, both on the bench and in flight.

It has been found that the two-stroke McCulloch engine, with a 3:1 reduction, swinging a 44" propeller, results in a quiet, powerful, powerplant. Its 22-24" pitch pusher gives a good rate of climb and cruise speed. Unless otherwise specified, assume all data figures using the McCulloch.

The plastic fuel cell, located between the pilot's legs, contains two gallons of fuel/oil mix. However; only about seven quarts are usable. Under normal climb/cruise conditions, this fuel supply will last well over one hour of flight, with plenty of reserve.

Assembly and Teardown

Because the Mitchell Wing B-10 can be assembled or torn down in less than 10 minutes, hauled on or behind the family auto to almost any place, operational costs are kept to a minimum (Figs. 6-1 through 6-4). You simply take it with you, assemble it, do your flying, take it down, and return home with it (Fig. 6-5). You escape the pocketbook drainers of tiedown fees, hangar rent and glider tow charges. Once groomed through practice, you can accomplish teardown or put-together in less than five minutes. Here's how simple it is:

Take the wing from your cartop by removing four Quick-Pins (Q-Pins) and two cable retainers; set the wing on any of the flying cages, and install the same four pins, mating the wing to the cage; take out the six Q-pins holding the folded wing portion, then reinsert the pins to keep the unfolded wing panels in position; attach the wingtip rudders with two more pins; insert the control stick using another pin; and you're ready to take to the heavens. Teardown at the end of the flying day is just as simple.

Walkaround Inspection

Begin the inspection at the front of the ship. Check the rudder controls going to the steerable nosewheel unit and to the rudder bar. Check all control connections.

Move to the right, looking over the leading edge for the obvious—chips, dents, holes and cracks. Recheck all the safeties

Fig. 6-1. Ford Ranchero pickup hauls B-10 easily.

on the three pins that hold the wing panels in position—leading edge, upper and lower spar pin fittings.

Walking to the wingtip rudder, check its retaining pin—be sure it's further safetied with tape. Make a visual inspection for damage.

Moving on around to the trailing edge, inspect the stabilator for security. Examine all pertinent retainer pins and bolts. Look over its leading edge—as it's a separate, free-flying airfoil—as you did with the main wing's leading edge. Look for paint, dope and fabric damage. At the engine end of the stabilator, observe the pin connections that go to the push/pull rod control activator.

Next, turn to the engine. Go over it for engine mount cracks, and generally give it an inspection that would accompany any

Fig. 6-2. B-10 cage ready for wings.

Fig. 6-3. B-10 cage with wing installed, outboard panels still folded.

powered aircraft engine. Firmly grasp the powerplant, and give it the old "heave-ho," making sure it's on solid and all fittings are within limits.

Inspect the wooden propeller for cracks, dents and chips. Scan the wood's grain for solidarity and smoothness. There are a number of bolts connecting the prop to hub. Make sure they're tight and that there are no cracks or flaws radiating from individual mount holes.

Next, move to the left stabilator, rudder and around to the leading edge, back to the cockpit area, looking for the same items just gone over on the right side.

At the cockpit area, scrutinize the four retaining pins that were installed during assembly to hold the cockpit or tricycle gear cage to the wing.

Fig. 6-4. Outer panels in flying position.

Fig. 6-5. Wings fold back for trailering and/or storage.

Startup and Taxi

Climbing aboard the low-slung undercarriage can be a mite tricky, but once you've got the knack, it's a real snap. Although the craft comes supplied with a padded seat, back support and safety belt, creature comforts are left to your personal modifications.

Sitting in the cage, operate the rudder pedals (Fig. 6-6). You can usually see a shadow of the rudders on the ground. Watch this shadow when applying rudder pressure to see if it moves correctly. Check the stick control or command bar for forward, aft and lateral movements. Overhead is a clear plastic window, allowing visual inspection of the inside wing area. Through this window, inspect the control/mixer box for components, position and operation.

Once you're satisfied with the walkaround and control checkout, make sure the switch is "off," and pull the compression release. Look to the fuel lever—it's wired into the "on" position with light wire. If you want to turn the fuel "off," the wire breaks easily.

With the fuel "on," pull the choke control, reach over your left shoulder with the right hand, and pull the engine through four or five times. Now, with the switch "on," choke "off," the throttle cracked about halfway and your feet planted firmly on the ground—so the machine doesn't lurch forward with fire up—give the self-disengaging starter pulley a tug. It should crank right over and catch.

As the engine fires to life, it should settle in the area of 4000 rpm. Let it warm up there for about 30 seconds or so—until its temperature stabilizes. Once this has been achieved, and with your feet still planted firmly on the turf, rev on up, momentarily, to 7500-8000 rpm, then back off to idle. You're now ready to taxi.

In learning to fly the Mitchell Wing, you should spend a lot of time taxiing—perhaps two hours, running it around the airport area, taxi strips and around/through ground obstacles. Get the feel of the controls, starting and stopping the engine, as well as the ship itself, and how to park it. In general, get to feel comfortable in all modes of ground handling.

Takeoff, Climb, and Landing

Inspect your remaining fuel supply, making sure you have at least half a tank. Now you're ready to try that first liftoff. Actuating the small lever-type throttle control on the left diagonal brace with the left thumb and forefinger, give it full power, and accelerate down the runway. Hold forward on the stick until 20-25 mph. It's going to want to fly, and *will*, if you don't have the stick in the forward position.

The craft picks up speed at an amazing pace, and rotation is almost immediate. Now ease the stick back. Remember, *ease* it back—don't pull-jerk back. If you do jerk it, the Mitchell will

Fig. 6-6. Sitting in B-10 cockpit, ready to go.

simply jump into the air. By easing it back, the B-10 will gently rise off the ground (Fig. 6-7).

Actual liftoff depends on a combination of different items—soft or hard surface, power/rpm, gross weight, etc. Using the McCulloch 101 engine, with a hard surface and revving to about 8000 rpm, liftoff occurs at less than 150 feet. With the same engine, same rpm, but a soft field, 200 feet is closer.

Once in the air, reduce power to idle. Let the ship glide on out, then bring it down for a landing and roll out. Now taxi back, and repeat the takeoff process. Each time, make the ground skimming a little higher and a little farther. Spend about an hour in this phase of ground skimming. This will give you a lot of takeoff and landing experience, with a minimum of danger to either you or your craft.

When you feel comfortable—regardless of amount of time spent in ground skimming—throttle to full power, accelerate down the runway, and lift off as usual. However; instead of reducing power as before, let the airspeed increase to 35 mph. Hold the ship level until this airspeed is reached.

Climbout book figure, using full power at 9000 rpm, is estimated at 400 fpm, but there's no way to check, since the machine carries no altimeter or rate of climb indicator (ROCI). A slower climb can be initiated, one that suits you and/or your needs.

Just like lifting the B-10 off the ground, ease the stick back to hold your established 35 mph. At the same time, throttle back to 7500 rpm. The Mitchell will be climbing out at 250-300 fpm. At this setting, the engine, aircraft and yourself won't be working hard at all. You'll find it very responsive in all control modes.

Airspeeds range from just above a stall at less than 25 mph to high cruises approaching 60. Using the craft as a powered glider, the length of flight, both time and distance, is strictly yours.

Climb it out to about 300-feet AGL straight ahead. Make a 90-degree turn, and go on up to 600 AGL. Make another 90-degree turn, climbing out to 800-900-feet AGL. At this point, all you want to do is continue on around the pattern and land.

On the downwind leg, opposite of where you want to touch down on the runway, simply reduce throttle and lower the nose to hold 35 mph. By adjusting the power and holding 35—the magic number—you can adjust your rate of descent. Hold this airspeed during the remainder of the downwind, base and onto final approach. In fact, hold 35 mph right down to flare out, two or three feet above the runway surface.

Once you've reached the flare altitude and are over the spot of intended landing, reduce power to idle, holding the aircraft off the ground as long as possible—as long as it'll still fly. You'll find touchdown about 20 mph. This phase will be like old home week after so much ground skimming practice (Figs. 6-8, 6-9).

Now simply taxi back, and go around the pattern again. Shoot plenty of pattern to landings—some full-stop, others touch-and-gos. It's here that you'll acquire the judgement, self confidence and feel of the aircraft.

Do your pattern work in calm conditions. Even though the Mitchell Wing B-10 has a high crosswind component, it's still an ultralight and subject to bouncing you around. Also, you want to make sure the CG is correct. You will, most probably, find a need to add weight to the nose, as these ships have a tendency to be tail heavy.

Cruise, Turns and Stalls

After climbing to altitude, drop the nose and let the bird accelerate to 45-50 mph, with the engine turning something less than 8000 rpm. That's about the extent of cruise. Watch engine temperature, and make rpm adjustments accordingly. From here on, you can adjust speed/rpm settings to accommodate your own desires.

Turns are accomplished with the use of wingtip rudders mostly, using very little bank. Naturally, the steeper the bank, the more back pressure on the command stick to keep the nose from falling through. The B-10, much like its newer brother ship, the U-2, has a tendency to drop its nose if back pressure is released on the stick.

Steep turns are like those with any normal aircraft. You have to use a lot of rudder with the B-10, however. Once you've tested turns in all modes, you'll find them no problem.

Fig. 6-7. Ground skimming.

Fig. 6-8. In the air and on the climb.

Approaching a stall, the 230-15 airfoil center section will always burble first. This is intentional for increased safety. The outer wing panels are still flying when the center section stalls. This prevents the machine from stalling/falling to one side as with most conventional birds.

Climb it up to about 1000-feet AGL, and start easing back on the stick. Reduce power to 4500-5000 rpm. Keep easing the stick back farther and farther—a very gradual process—until it stalls. The B-10 will shudder, and its nose will fall through at about 20-22 mph.

As soon as it shudders and falls through, simply release back pressure on the stick, thus dropping the nose, and the ship will recover almost instantly.

Stall effects are very gentle, with almost no break, unless intentionally forced into one with power or zoomed without power. Recovery is quick and clean, with the loss or relatively little altitude. I would not recommend it, but I've observed masters with the B-10 doing stalls and recoveries within 50 feet of the ground.

Spins aren't recommended in either of the Mitchell aircraft, although they are perfectly capable. However; if you inadvertently get into one, recovery sequence is as follows: release back pressure on the stick or pop it forward momentarily, apply opposite rudder, and the aircraft will recover almost immediately.

Inflight Starts and Descents

The B-10 has a glide ratio of more than 16:1, meaning that for every foot of drop in altitude, the machine will glide 16 feet forward. That puts the B-10 in the same class as some pure gliders.

Maximum glide is achieved at 36 mph, with minimum sink rate at 26.

Inflight starts of the McCulloch engine are no problem, either. Having the capability of inflight starts allows you to cut the engine and play in thermals or soar. On restart, most probably, the engine will have cooled. Leave the switch "off," and pull it through four or five times with the choke "on," just like on a ground start. Now, with the choke "off," and the switch "on," pull the compression release a couple of times, and simply fire up.

Normally, you've got the stick in your right hand, making it somewhat difficult to reach over your left shoulder for the starter pully. However, it's fairly easy to hold the stick with your right knee while pulling the rope through with your right hand. It'll normally start right off on the first pull through.

Because the craft is a pusher, expect a drop in the nose upon engine firing. It's not much of a drop, but enough to surprise you, if not expected. There are a number of systems used to keep from having to use the "knee method" on the command bar, two being bungee chords and, of course, spring attachments. In either or these two modes, there's no problem flying it hands off.

As previously stated, wingtip rudders on both Mitchells can be used as airbrakes or for steep descents as spoilers, but there's seldom a need to use them for those purposes. With the power off, it's only a matter of putting the nose down, and letting the craft accelerate to 55-60 mph.

Normal descents are made at 30-35 mph, about the same as climb speeds. Approaches are made using your own judgement. You have to allow for winds, watch for crosswinds, come on in, do your flare out, holding it off until it has quit flying. Then taxi back.

When the flying is done, and you've taxied to the car for hauling home, you're ready for shutdown. Simply pull the com-

Fig. 6-9. Flare for landing.

Table 6-1. Mitchell Wing B-10 Performance and Specifications.

POWER

Engine Type: McCulloch
Number of Cylinders: 1
Horsepower: 10@8000 rpm
Propeller: Wood—hand carved
Propeller Diameter: 44″
Propeller Type: Fixed pitch
Aviation Fuel Grade: Fuel/oil mix
Power Loading: 3.5 lbs/hp

SPECIFICATIONS

Wingspan: 34 ft
Wing Area: 136 sq. ft.
Length: 8 ft
Height: 4 ft
Seating Capacity: 1
Maximum Gross Weight: 401 lbs
Empty Weight: 140 lbs
Usable Weight: 261 lbs
Fuel Capacity: Standard tank—2 gal (7 qts usable)

PERFORMANCE

Cruise: 55-60 mph
Stall Speed: 22 mph
Takeoff Distance: 200 ft
Landing Distance: 95 ft
Best Rate of Climb: 400 fpm plus
Glide Ratio: 16:1 plus
Service Ceiling: 12,000 ft
Fuel Consumption: 1 gph

pression release. If this doesn't shut off the engine, hold the compression release and turn it "off" with the ignition switch. It'll normally shut down with the compression release, though.

Weights

With the basic wing weighing in at only 70 pounds, you can readily see why it's called an ultralight. Using Mitchell's specifications for a pilot of up to 250 pounds, gross is no problem in any mode of operation. For the foot-launched module, add another 10 pounds for the hang cage, and you have a leap-off weight of 330 pounds. Hang on the tricycle gear for 35 pounds, including engine at 35, and 11 pounds for seven quarts of go-power, and you're looking at 401 pounds lift off.

Building the Mitchell Super Wing U-2

7

Construction of the Mitchell U-2 begins with the same preliminaries as for the B-10: inventory of your materials and hardware, rounding up your tools (the same ones mentioned in Chapter 4), and familiarizing yourself with the assembly manual and plans.

Rib Construction

Lay the rib plan onto a piece of ¼″ plywood or a work bench, much as with the B-10 construction. Cover the rib plan with a clear plastic overlay, remove the wrinkles, and staple along its perimeter.

Take a length of ¼″ × ¼″ stock. Holding it by one end, wave it up and down to detect any flaws in the wood. If it breaks, take a longer piece and repeat the procedure. Keep repeating until satisfied you're working with a sound length. Don't discard any wood shorter than 6″. Remember, go easy—you're not trying to break it, but only detect flaws that aren't visible, such as knots or pitch pockets that take away from its strength. Repeat this procedure with every piece of wood selected.

Choose a piece of wood for the rib #12 top capstrip. Holding the wood's squared edge flush on the blueprint—"rear face of spar" line, and below the rib #12 line—nail above and below the stock to hold it in exact position. Follow the #12 line back—forming the wood and tacking nails until you reach the tip. Don't squeeze the wood too tight—just enough to hold it in position.

Upon reaching the tip, check that you've nailed it sufficiently every few inches. With a pencil, mark the vertical cut-off and the chord line cut-off.

Gently remove the capstrip from your newly formed jig for cutting to size. A razor saw works well, if a power saw isn't available. After cutting, replace the capstrip in the jig, and check for positioning.

Choose a suitable piece for the lower capstrip, and repeat the nailing process. The lower capstrip goes above the #12 line. Cut the vertical and chord line points, as with the upper, and sand to fit.

At this point, you should be able to barely see the blue outline of the #12 rib to the outside of your work. The triangular marks on plan #1 are the intersections of the vertical and diagonal truss members and capstrip.

Starting at the "rear face of spar" line, fit a vertical truss just to the right—so the line is barely visible. Cut it a bit oversize, then sand it to fit.

Be careful not to round the wood's edges. Once in position, place a weight on each truss member, holding it in place.

Each rib has a different spar face. The first diagonal truss goes from the bottom to the top. Check detail drawings at the top right of blueprint #1 for an illustration of the joints required.

Position all diagonal trusses, making sure that you haven't overlooked any intersect points. Witness mark all truss members. Now, with scissors, cut, fit and sand gusset plates and trailing edge side gusset plates. Witness mark these gusset plates, also.

Mix 10cc of the laminating resin to 2cc of hardener. This is a 5:1 mixture. Stir and mix thoroughly. Then, with a brush, dab the epoxy mix onto the tips of the capstrips at the trailing edge first, then onto both ends of the vertical member. Join the glued joints, and replace the weights. Continue with the diagonals, replacing and reweighting them as you go, using the witness marks as your guide (see bonding section).

Upon completion, glue the witness-marked gusset plates to the proper joint. Carefully staple gussets to the ribs. A word on stapling—don't shoot the staples directly in line with gusset-plate wood grain. Offset the staple as much as possible, to avoid wood splitting. It may be wise to try a few on some scrap wood first. Four staples per gusset is sufficient, except on the trailing edge gusset plates and the overlays. These will take stapling every ¼". Mark each rib, when completed, to indicate its top and rib number.

With the epoxy still wet, gently remove the rib from its jig, and turn it upside down. Install gusset plates on this side, also, using the above procedure. If there is any separation between truss members, squeeze them together while stapling. Unless you're very good, consider this first rib practice.

The most important ribs are #1, #7, #8, #10 and #12. Because of their importance, do them last. You're going to need a pair of each, except for four each of the #1 and #7 ribs. The #1 and #7 ribs are made of the ⅜″ × ⅜″ stock, with 2″ × 2″ gusset plates cut from scrap.

On the four #7 ribs, a hinge block is installed three-quarters of the way back. A detail at the bottom of blueprint #1 indicates its

Fig. 7-1. 2¼″ overlays.

position, along with two triangles on the rib plan itself. The 2 ¼″ plywood overlays go on both sides and are angled according to plan (Fig. 7-1).

In the outboard ribs #7, #8, #10 and #12, a gusset block is installed between the capstrips for the stabilator hinges. Again, fit this carefully. Glue into position with gusset plates on both sides. This is detailed at the top-right of blueprint #1.

All four #7 ribs have the first vertical member doubled. Apply glue to the mating surfaces. Two of the #1 ribs have engine mount support blocks—⅜″ × 1″—installed in the same fashion as the hinge blocks on #7. The other two have the same diagonal that's typical of all ribs.

Once you've glued, gusseted and stapled all ribs, set them aside. You'll be coming back to them after the epoxy dries to remove the staples.

Spar Assembly

Take the outboard spar caps, and, one by one, sight down each side of them to identify which is the straight edge. Mark each spar cap accordingly for future reference.

On a 1″ thick plywood sheet, draw a straight line down one of its 8′ sides, about 2″ from the edge. Nail small wood blocks—⅝″ thickness, or less—on the outside of the line every few inches. These blocks can't exceed ⅝″-thickness, or they'll interfere with the spar web bonding. They hold the top spar cap straight during assembly, so work carefully.

On one end of the straight line, draw a perpendicular line about 1″ from the edge of the plywood. Make this line about 10 ½″ long. This line indicates the outboard face of rib #7. It also determines whether it's a right or left spar.

Foam verticals are to be placed perpendicular to the top spar cap. For this reason, draw lines every 4 ½″ from the #7 rib to determine the positions of the other ribs. Number out 21 spaces from rib #7.

With the straight edge to the outside, line up the large edge of the spar cap to the line representing the inboard face of rib #7. Block to hold in position, being certain not to interfere with the foam verticals which go on each of the drawn lines you've made. Leave the very end of the spar cap from rib #11C free of blocks.

From the intersection of the top of the spar cap and the #7 rib line, measure down 8 ¾″ and make a mark. Now, from the top of the spar cap at rib #12, measure down 4 9/16″ and make another mark. With a straight edge, connect the two marks, continuing the line out to an equal distance with the top spar cap. Block the lower spar cap in position, making sure the taper is facing the top spar cap.

Now remove the spar caps, and cover the jig with plastic or wax paper for a good bond release. At the inboard end of the spar, in place of ribs #7 and #7A, place ¼″ wood verticals. Check page #3 of the drawings.

Laminating resin can be used throughout spar construction. Fix, and glue the 1″ × 1″ foam verticals on the lines previously drawn, working out from rib #7. After the glue is dry, saw and sand the foam flush with the capstrips, but don't overdo it. Install blocks at the tip, as per plan.

Skin the spar with a #3 and #4 1mm birch plywood web, by squeegeeing laminating resin on the entire surface of the web, foam verticals and the edges of the capstrips—about one cup will do. Clamp, weight, and staple the web. Make sure you have a good, sound bond. Glue, and place the 2″ wide 1mm plywood over the splice.

After drying, trim overhanging plywood, and sand off excess glue. Glue and staple the spar web doubler at the rib #7 uncovered end. Let it dry, then trim excess. Strip the jig so that only the initial row of blocks remain.

To make a left wing, lay it out on the opposite side of your straight line, and repeat the entire procedure.

Inboard Spars

First thing in building the inboard spars is to put together the four laminated bows required to make the 12-degree swing (sweepback) in both right and left inboard spars. Details are found on pages #6 and #7 of the blueprints, and put together much as the bow section of the B-10.

Take a plywood sheet, and lay it on the floor. Come down from the center-top of the plywood sheet a few inches, and make a mark. Now, measure down and off the plywood sheet 72″ from this mark. This 72″ is the radius for the semicircle drawn at the mark on the plywood, which indicates the curve of the 12-degree sweepback.

Now, on both sides of the centerline of this semicircle, measure out 6″, and make a mark. These two marks are where the scarfed ends of the spars will come to when laminating the bow to the spars.

To make the actual bow, you need four ¼″ pieces, or eight ⅛″ pieces, or five 7/32″ pieces—as long as their combined width will come to right at 1″. These pieces should be 36″ in length when bonded to the spars, so, initially, start off with something on the order of 44″-45″. Simply laminate these individual layers together, following the outlined semicircle for 6″ on either side of the reference centerline. Use plastic or wax paper, along with bonding blocks, to insure both a tight glue joint and bond release after the 9410 structural adhesive has dried. Scarf the outside edges of the laminated bow for 12″ from the final measured and cut-to-form ends.

On a flat table, adjust the #1 and #2 spar caps so that each scarfed end mates with the bow at the 6″ mark. With the spar cap firmly in place, mark the bow for cutting.

Laminate with the 9410 adhesive. Use clamping blocks with plastic and six clamps per joint. The spar has a 12-degree sweep that makes assembly somewhat more difficult. An extension to the table must be made now—one that projects upward at 12 degrees for at least 62″, and one that's 12″ wide.

Secure the top spar cap—#1 and #2—parallel to one side of the table, with the centerline on the edge of the extension. Secure the lower spar cap to the table or plywood sheet. This is accomplished 8 ¾″ down from the top edge of the upper spar cap to the bottom edge of the lower spar cap at rib #7.

Lay out the rib stations down to the #1 rib. They're at 18″ centers, down to rib #4. Rib #3C through #1 as per plan #6.

From the top edge of the upper spar cap, at rib #1, measure down 13 ¾″ to the bottom edge of the lower spar cap. Scrap wood tacked from upper to lower cap is helpful here.

Divide each rib bay into four stations, with the exception of rib bay #1. It's divided into only three. At rib #7, and the first inboard station, wood verticals replace the 1″ × 1 ¼″ foam spacers which are located at every station down to rib #1. There are also one wood vertical for rib #0, adjacent to rib #1, and two wood verticals for the landing gear.

These wood verticals, plus foam spacers, are to be glued in place before the web is applied. Install the foam spacers so that they protrude above the spar caps. This excess is trimmed flush after they're glued in place.

The web is applied with a thorough coat of epoxy and a minimum of two rows of staples shot into the spar caps. Weights are needed to attain a good bond between foam and plywood. It takes three sections of plywood to complete the spar. They are butt-spliced on convenient foam verticals, with 2″-wide doublers cut from excess stock.

After the web has dried, a web doubler is applied on the opposite side and the #7 end.

Wing Attach Fittings

In the middle of the blueprints on page #3 is the detail on the wing hinge point (Figs. 7-2 and 7-3). Eight metal fittings cut from these patterns are required per side.

At the #7 end of the inboard spar, wood shims are laminated with the 9410 adhesive on both sides (Fig. 7-4). They should overhang the fittings by at least ¼″—spar web sides only.

Next, nail 2″ × 4″ blocks on one end of the plywood sheets. This is done to elevate the inboard spar. Make sure the foam spacers are facing up. Now center the #7 end of the spar on the plywood, and secure.

With the straight edge, draw a horizontal line extending out from the top of the spar end. From a point ⅞″ from the edge of the spar, draw another line, this one 6 degrees up. Continue this line out to the end of the plywood. This is the 6 degrees of dihedral to be set into the outboard spar.

Block up the outboard spar, making sure the top is flush to the drawn line and the rib #7 end—approximately 1 ¼″ from the top of the inboard spar. Secure to jig, but leave room for adjustment. Clamp the .125 metal fittings in place—top and bottom—and check that the shims are going to allow clearance.

Put a 3/16″ bolt through the fittings, and adjust the inboard fittings according to the measurements on the blueprints (Fig. 7-5).

Check the shims to insure enough clearance—1/64″ to 1/32″. Sand or laminate scrap birch plywood to fit. Measurements of the lower fittings could vary slightly from 1 ¼″. Double check the angle between the two spars at the top capstrip. Mark the fittings and their locations. Drill the two 3/16″ holes in each of the four assemblies. Remove the fittings, and deburr the drill holes. Sand and shape the inside surface of the metal to the edge of the spar. It's important not to touch the surface after sanding to insure sound bonding. Bond immediately after sanding.

Outboard "D" Tube

Trace the spar onto a plywood sheet, and cut ¼″ inside the line so that the edges of the spar hang over. Drill holes in the plywood, and mortise so the attach fittings aren't distorting the spar. Nail a 2″ × 4″ block on the edge, flush with the long sides. Drive nails even with the foam, every 4 ½″ through each of the blocks, with approximately ½″ of nailhead above the surface. These nails are positioned for the surgical tubing that holds the skin in place while gluing.

Bond the reinforcement blocks to the end of the leading edge with 9410 adhesive. After they dry, sand the mating surface of the leading edge attach fitting, as shown on page #3 of the blueprints. Bond with 9410, according to the detail at the top right on page #10 of the blueprints.

Secure the spar to the jig, making sure the nails are even with the foam. The foam must be facing up. Take one of the four ¼″ plywood #7 nose ribs. The #7 plywood rib has two birch doublers

Fig. 7-2. Wing attach hinge (4 of 8) inboard #7 rib attach.

Fig. 7-3. Wing attach fittings.

glued to both sides of the leading edge where the drag strip attaches. Staples can be used for pressure. Also, the rib has two ¼″ × ¼″ strips positioned vertically for rigidity. Notch the bottom of the rib to accommodate the wing fittings and shims. Determine which is the top of the spar by referring to the detail at the top of page #3 of the prints. Sand the rib down to conform to this pattern.

The wood rib goes flush with the end of the spar. You must notch the ends of the rib to clear the attach fittings and the shims. Also, the correct leading edge must be fitted—½″ outboard; ⅝″ inboard, according to plan.

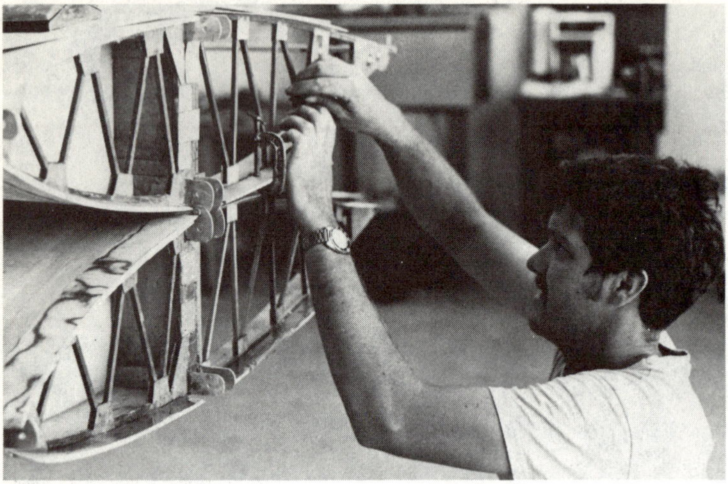

Fig. 7-4. Gluing wood shims.

Now, with 9410 adhesive, glue the #7 rib to the spar. Use small, 1mm plywood gussets, and staple to hold to the spar. Run a thin fillet of adhesive between the spar, fittings and rib.

Using a framing square, hold the rib 90 degrees to the spar, and brace until the glue dries. Check that the camber is on top. Do the same to the #12 rib, assuring that it's positioned to correspond to the #7 rib and top of the spar. Label the white foam ribs #1 (the largest) through #19 (the smallest), on each set of spars.

Now glue the ribs to the spar and spacer, either wiring or stapling them in place, making sure to keep the leading edge straight. Check for sequence and top of camber. After all of the ribs are in place, hotwire the notch for the leading edge.

When the glue sets, epoxy the leading edges to them, making certain they're perpendicular to the spar. It's important that the leading edge is bonded to the #7 and #12 ribs with the 9410 adhesive. Take care that the leading edge and reinforcement blocks are flush with the outside edge of the #7 rib, with the fitting sandwiched in between.

Hotwire the ribs to approximate shape by using scraps of plywood stapled to the #7 and #12 ribs. Get a long, straight 2″ × 4″, and staple sheets of heavy grit sandpaper on the face for 90″. Leave hand holds on each end. On both ends, also, staple a 1mm plywood strip. These 1mm plywood strips are used to guide the 2″ × 4″, following the contour shape of the #7 and #12 ribs, as you sand the foam ribs to shape.

Sand the ribs slowly, making sure to keep the sanding board in the same position on the end ribs. Sand gently, so that the ribs aren't broken loose in the process. Before you finish sanding, round the leading edge spar to conform to the airfoil shape.

Check that total contact will be made with the skin, especially where the spar and foam meet. There are two skins per spar. Make sure you choose the proper skins. They're cut oversize to permit adjustment.

Starting at wood rib #12, lay the first skin over the foam, and strap it down with surgical tubing. Be certain there's overhang on both the top and bottom of the spar. There also needs to be a certain amount of overhang on the end of the spar. The other edge of the skin is on the rib. Mark for upper fittings, and trim.

With the first skin still on the spar, lay the second skin over the foam, and adjust. The skins need to overlap by at least 1″. It might be helpful to tack two nails through each skin into the leading edge to align them. This is done before the glue is applied.

Surgical tubing must be strapped over the skin at every rib station. Draw this tubing tight enough to insure a good bond between the skin and the foam. If your geographic area is particularly dry, moisten the skin while in place, and let dry before gluing. It's important that everything be ready before any glue is mixed.

Once you're ready, work up about ½ cup of the laminating resin per skin. Paint each rib edge, and the entire spar cap, up to

Fig. 7-5. Wing attach fittings laminated and bolted on wing spar.

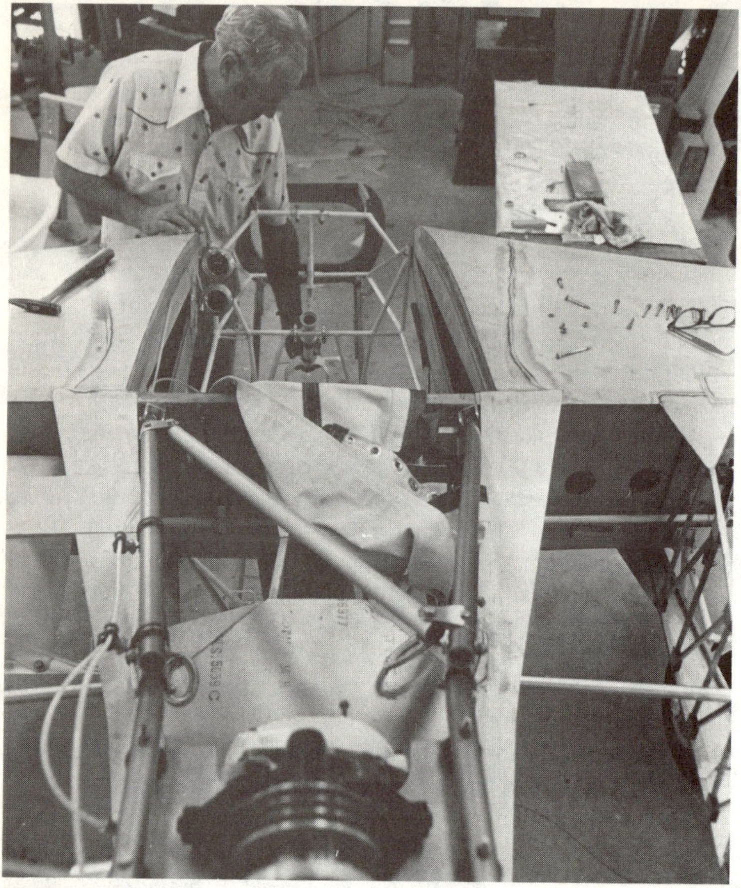

Fig. 7-6. Wing skins butt-spliced into cockpit area.

the overlap. Spread an even coat over the skin, and lay it in position over the foam, using the nails as guides. Strap in place with the tubing at each rib. Staple the skin to the spar every ½″, except for the 1″ overlap. You might have to put some staples in the #12 rib, if you can't get the skin to shape properly otherwise.

Repeat the procedure for the next skin, taking care enough glue is applied to the 1″ overlap.

Inboard "D" Tube

A #0 nose rib must be constructed from the two types of laminations provided. The technique is the same as that of the trailing edge ribs. Also, a fixture to hold the spar must be made by

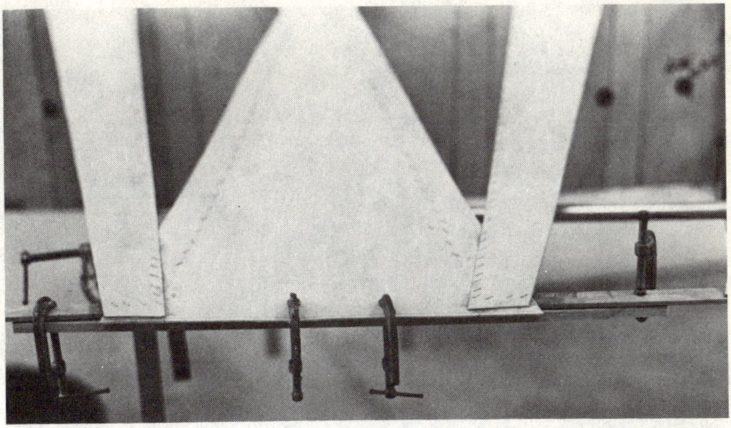

Fig. 7-7. Trailing edge bonding with gusset, as described on page 6 of blueprints.

cutting a male image of the spar from plywood, reinforced with 2″ × 4″s.

Sand the mating surface of the metal fittings, and bond the rib to the spar with the 9410 aerobond. Two 1mm gussets that are glued to the spar end and rib helps to hold the rib in position while

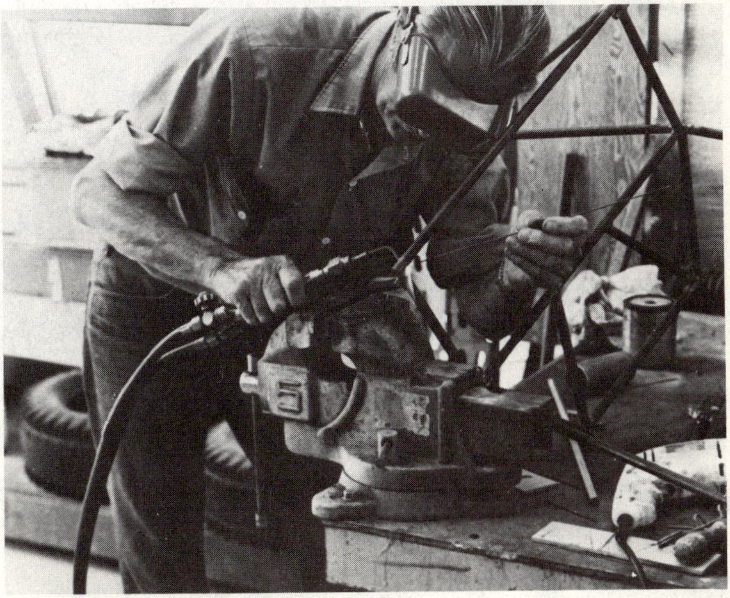

Fig. 7-8. Welding tubular cockpit frame.

Fig. 7-9. Measuring skid.

drying. Be sure the top of the camber is matching the top of, and is 90 degrees to, the spar.

The #0 rib mounts to the wood vertical, the same way, using the 1mm gussets, also. This must be rigged leaning inboard 2 degrees. Staple scraps to hold it in position. It's approximately 113″ from the leading edge of rib #7 to that of the #0 rib. Foam ribs #7 through #4 mount the same as the outboard. Rib #3C through

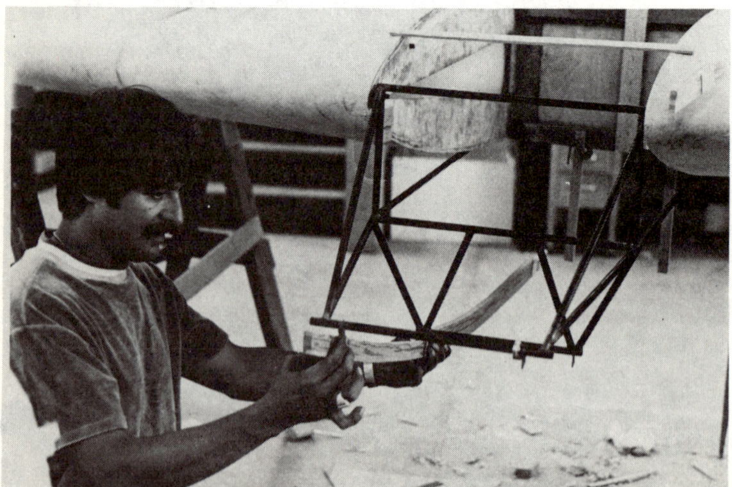

Fig. 7-10. Adjusting skid.

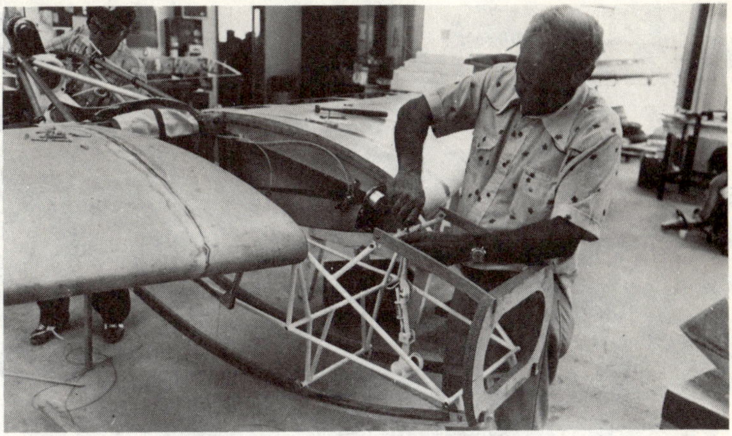

Fig. 7-11. Adjusting nosewheel retracting mechanism.

#1 have a sharp bevel that must be sanded down to appropriate size.

After gluing them in place, use the hot wire for the leading edge and sides. The skins must be butt-spliced on the inside with a 2″ doubler (Fig. 7-6). See details "A" and "B" on page #6 of the blueprints. And don't forget to make a left and a right (Fig. 7-7).

Cockpit Area

Basically, what you've got is a lot of chrome moly tubing (Fig. 7-8) and two laminated skids. You eventually have to laminate some additional strips to the bottom of the skids (Fig. 7-9). These

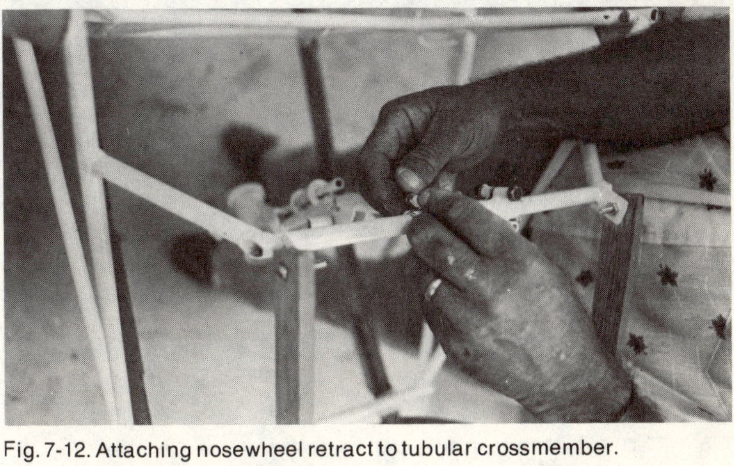

Fig. 7-12. Attaching nosewheel retract to tubular crossmember.

Fig. 7-13. Skinned cockpit area. Notice forward bulkhead and rudder pedals.

strips will give you more surface area for the birch plywood cockpit skins.

After beveling the skins down to the proper angle, you can either staple and glue them into position as you go, or you can staple in a mockup of the skin that comes into the armpit of the wing from the skid. Cut and fit until satisfied you're not going to need too much filling when finished. There's no way of putting it together

Fig. 7-14. Main spar running at rear of cockpit structure.

that doesn't require a certain amount of putty fill at the cockpit, leading edges and seams. The idea is not to end up needing too much.

From the forward bulkhead, right back to underneath the main spar is plywood. Generally, you just have to let the plywood take its own shape, with a certain amount of stringers laminated to the inside to keep the plywood from buckling (Fig. 7-10).

From the spar to the trailing edge of the wing, it's best just to cover with fabric. Patch the whole thing with Dacron fabric, and heat-shrink it until it takes the proper shape. Then you can either go back inside with more stringers, or leave it the way it is, and dope it. See page #5 of blueprints and Figs. 7-11 through 7-15 the tubular cockpit frame details.

Stabilator

Wet all six skins on a 2″ swath down their centers, where they're going to take the bend over the nose. This procedure is a duplicate of that on the B-10 stabilator. After a good soak, try to work them gently into the leading edge shape. Flex the plywood with your hands, and, again, don't rush it, as these skins crack rather easily.

When you've got them close to the desired shape, clamp or band them down until dry. Two parallel 2″ × 4″s, about 2″ apart, can also be used—eventually closing them to about ¼″. It might

Fig. 7-15. Painting gear mount wheel well.

Fig. 7-16. Finished stabilator hanging on wing before covering.

take two or three soakings before getting them right. However, with a little patience, they form nicely.

While the skins are drying, bevel and bond the reinforcement blocks on the aft side of the spar with the 9410 adhesive. After the glue dries, sand off any excess, being careful not to round the spar edges. See details on page #4 of the blueprints.

Glue the nose ribs in position. Be sure to make a right and left. This is accomplished by reversing the top of the ribs. The ribs can be glued in place with either laminating resin or the 5-minute epoxy. Bond wood ribs with the 9410. The skins are glued in place with the laminating resin.

There are various ways to get the skin down tight on the ribs. One way is to make female cut-outs of the nose ribs from ¼″ plywood. You should have these left over from cutting male forms for skin shaping. Press them down over the skin, and staple or clamp them with the 1mm plywood to the spar. Be sure to coat with resin the entire inside surface of the skin, the face and sides of the spar and each rib. Make certain a good bond is maintained with the spar.

Make the trailing edge ribs, using the fixture outlined on page #1 of the blueprints. The trailing edge ribs are glued in place according to plan. It's necessary to trim the excess plywood flush with the spar, before bonding the ribs. Put these on after the stabilator is mated to the wing to allow access to the nuts holding the fittings.

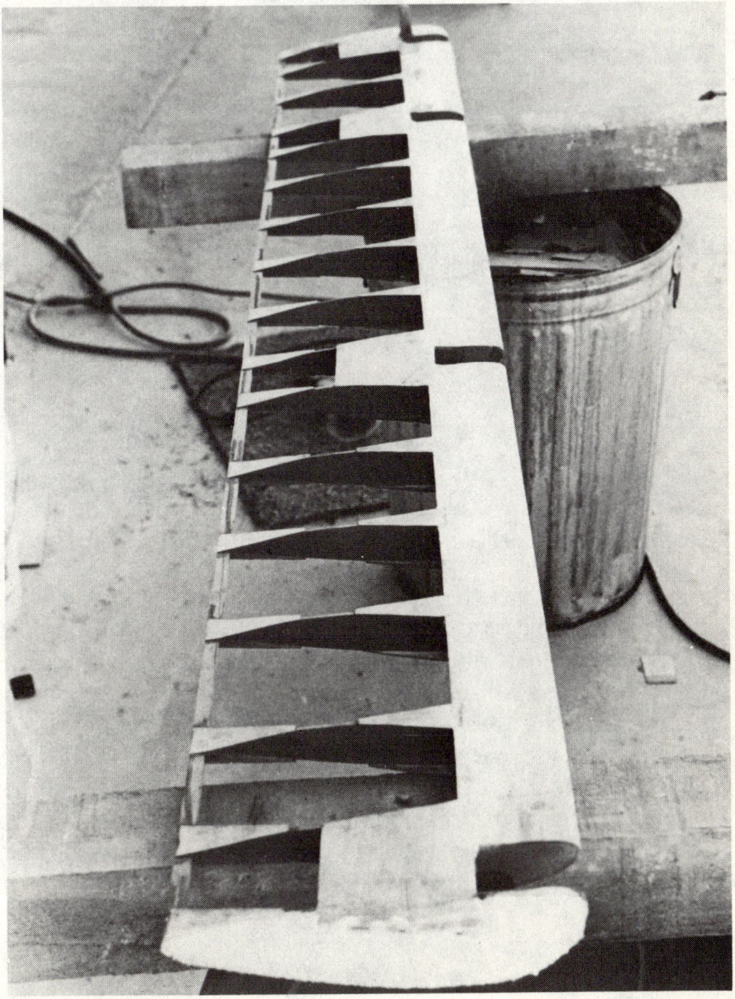

Fig. 7-17. Overall view of stabilator, trailing ribs and gussets.

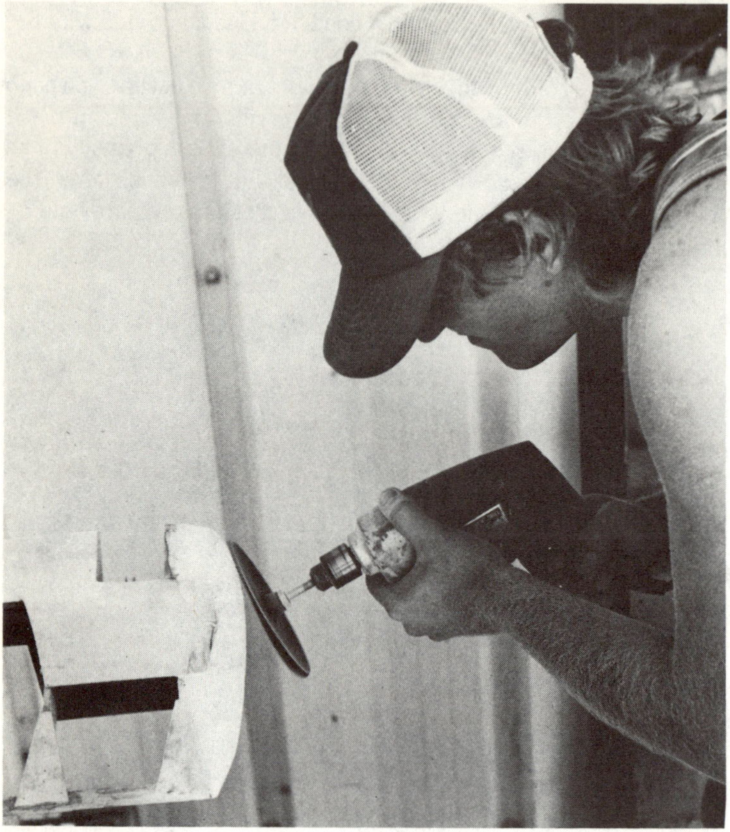

Fig. 7-18. Sanding faired stabilator ends.

Remember, make a right and left washout. The trailing edge can be put on as you bond the rib caps, or put on last—It must be kept straight, either way. Clamping the rib gussets to the trailing edge is preferred. Trim excess plywood, and mate to the wing.

Carefully drill the spar, allowing room for movement between the fittings and the bracket. Clean and bond the fittings to the spar with structural adhesive, and bolt with the 3/16″ × 1″ bolts. Bond the large rib caps in position, and trim excess. Sand all rough edges prior to covering. Use scrap fabric from the main wing to cover the stabilators (Figs. 7-16 through 7-20).

Wingtip Rudders

First thing on the rudder assembly is to cut and shape the individual rudder ribs, fore and aft components, as shown on page

#2 of the blueprints. Next, drill all the lightening holes. Now lay a large piece of plastic on your work bench—large enough to accompany a full-sized wingtip rudder.

Clamp the two spars together, the clamps being situated so as not to interfere with either the ribs or the ⅞″ aluminum post. The clamping simply keeps both the fore and aft components rigid as you build.

Position the spars on the plastic work surface with nails in a similar manner as when making the ribs. Make sure they're positioned perpendicular to the table. The same holds true with the fore and aft ribs, ascertaining they are on the same plane.

Apply the aluminum tube through the ⅞″ hole, and make sure it's parallel to the spar. Mark the area of contact with the ribs, and

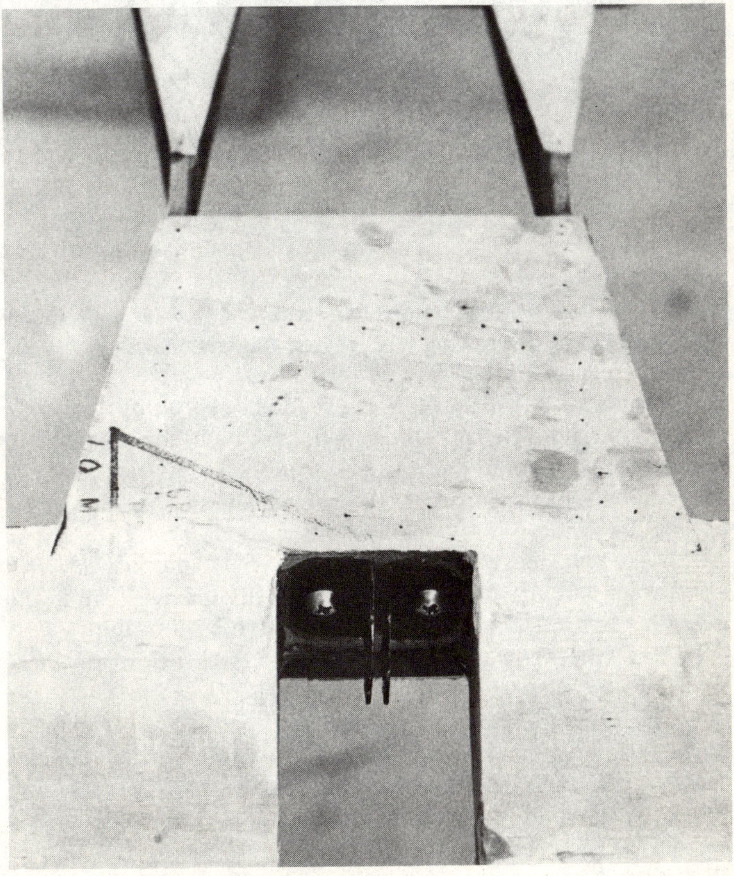

Fig. 7-19. Stabilator wing bracket.

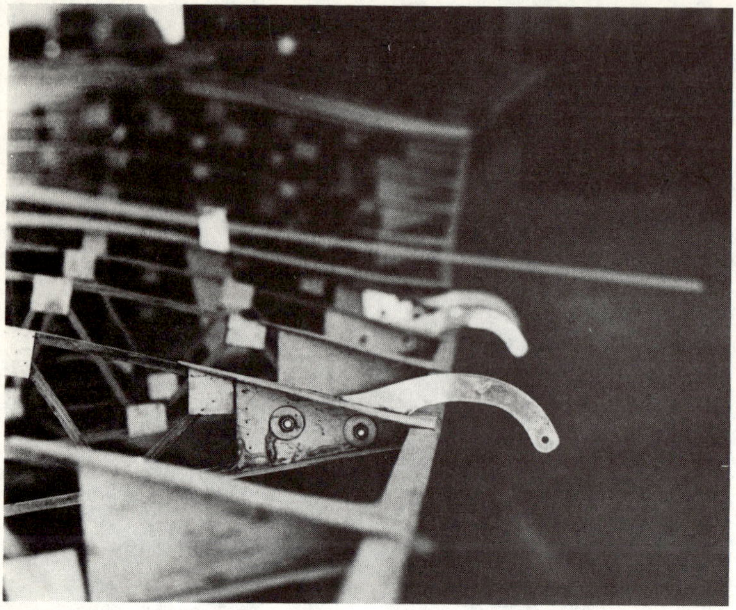

Fig. 7-20. Stabilator wing attach fittings.

sand. Clean the aluminum with MEK, and bond with 9410 adhesive.

Bond the 1 ¼″ birch plywood strip that ties the ribs to the individual spars. Also, bond the leading and trailing edge with the appropriate gussets.

Remove from the jig while still wet, and repeat the procedure on the other side. Be careful not to disturb their position. Before the glue sets, separate the spars. At this point, you want to attach the hinge mechanism—the three individual hinges and the return spring.

Make a full-length bottom rib that mounts to the spar tip. Drill a ⅞″ hole, and slip it onto the base of the rudder. Weld up the mounting brackets, and slide them onto the base of the rudder. Now clamp the whole assembly at the spar tip, at the proper angle. Drill the holes through the brackets at the spar tip.

The 12-degree outward angle is set by the lock pin hole and can be done any time after mounting. It's best to position the rudder assembly before the trailing edge ribs are attached to the spar.

After the rudder is in place, and has been set at the 12-degree outward angle, fare in the entire assembly with foam and smooth finish (Fig. 7-21).

Landing Gear

First thing is to copy the truss assembly that holds the wheel itself—getting the 5/16″ tubing in the proper location for the hinge mechanism. Attach the three channels to the rear of the inboard spar to the rear of the ¾″ bolts that are protruding out—two on the bottom and one at the top. Bolt the gear assembly attached to the lower fittings, and check for ease of movement. See landing gear makeup on pages #7 and #9 of the blueprints.

Fit the broken leg to the attach points—upper spar cap channel and flange fitting on the gear itself (Fig. 7-22). Position the bellcrank to the broken leg, and make certain it's not going to interfere with the spar throughout the movement of the landing gear.

A torque tube runs out from behind the cockpit area, through three brackets, to the bellcrank on the broken leg of the gear. Position the bellcrank and activating arm from the torque tube

Fig. 7-21. Wingtip stabilizer used in conjunction with spoilers instead of stabilators on experimental model.

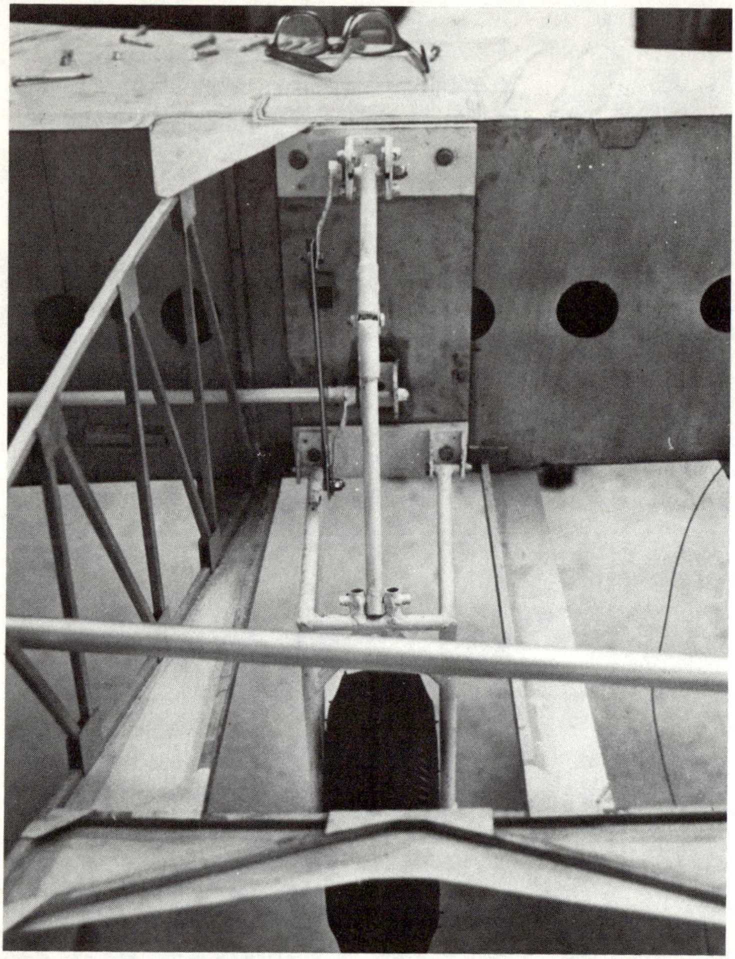

Fig. 7-22. Landing gear. Notice broken leg joint.

through the brackets. Bond the brackets to the spar, taking care that the bellcrank isn't hitting the back face of the spar at any time.

Adjust the large cam lever attached to the torque tube, while the gear is in the "up" position and the handle is full forward. Check for ease of movement throughout full gear travel.

Canopy

The canopy is started from scratch with the naked, premolded plexiglass (Fig. 7-23). It has a 23″ surrounding flange. You can set the whole thing on the cockpit opening and start blocking it.

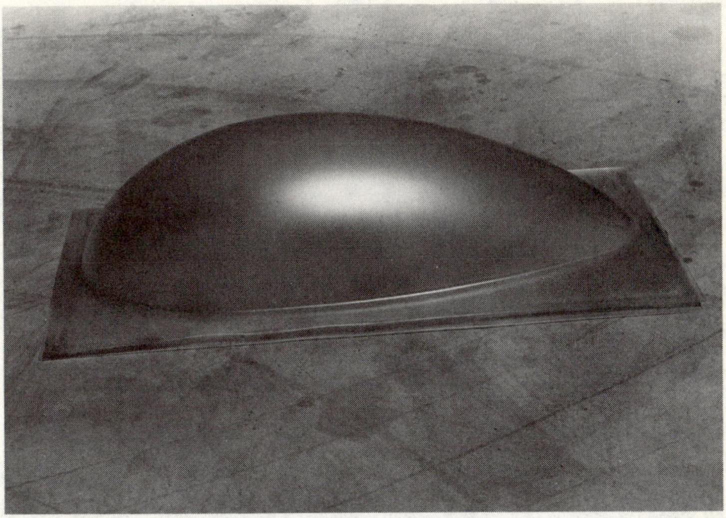

Fig. 7-23. Canopy plexiglass.

Eventually, you'll want to attach some wood for the locking pins and latches, as well as give it some rigidity.

Your first step in making the bubble canopy is to get into the ship. It's best to have someone assist in positioning the plexiglass bubble as to height, with you blocking it in position. Trimming may

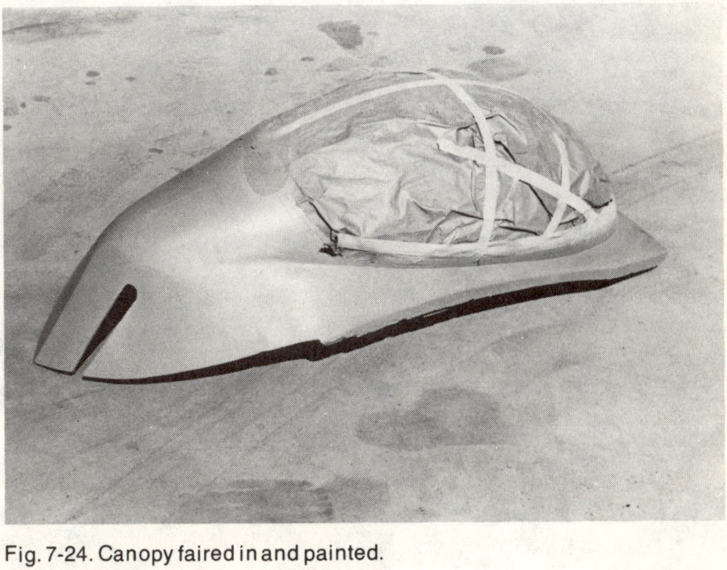

Fig. 7-24. Canopy faired in and painted.

Fig. 7-25. Completed canopy.

be necessary at the front of the canopy to get it down to the proper width—it's about 19″ at this point.

Next, bond ¼″ wood to the perimeter of the plexiglass, both above and below the flange area—In other words, sandwiching the

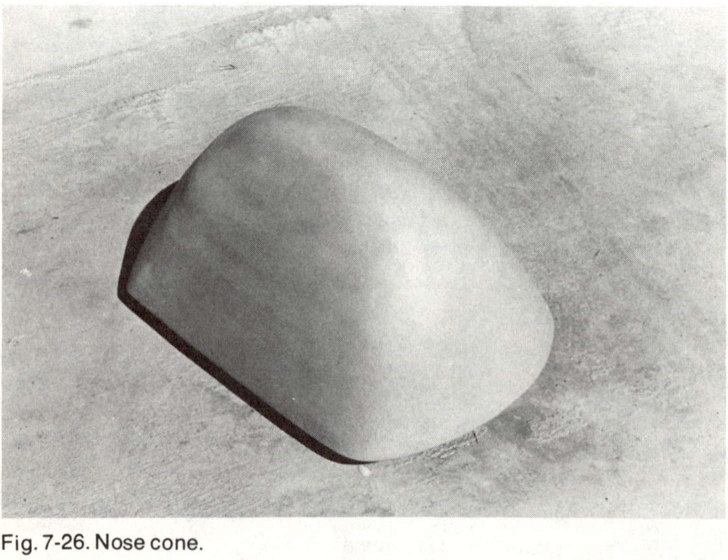

Fig. 7-26. Nose cone.

Fig. 7-27. U-2—first experimental model.

flange between the wood. Extra wood is needed at the front for the locking pins and midway down the sides for the inside latches.

Install the pins and latches. About the only criteria here is to make sure the canopy can be latched tightly, and the pins are secured firmly in the front.

After you're satisfied that the whole canopy structure is sound and blocked in the correct position for you, the whole canopy flange area is fared with foam and epoxy. Simply mask off the bubble portion and paint (Figs. 7-24 and 7-25).

Fig. 7-28. Spoiler on experimental model.

Nose Cone

Building the nose cone can either be a very easy and fast operation, or it can be one of the longest, most strung-out affairs of building the U-2.

The nose cone is basically only a stack of 10 individual layers of 1″ thick foam, shaped with a surform to some reasonably attractive-looking nose shape (Fig. 7-26). Bond the layers; then, with a Surform, shape the nose cone as a continuation of the cockpit frontal lines. It's either puttied over with Microbaloon or Stitts plastic filler.

Whichever of the above is used, there must be a complete covering. If not, when applying the dope, it'll eat holes or pockets in the foam. If that's the case, you'll more than likely have to fill in and redo, or start over again from scratch. You'll run into this type of problem again with the nitrate dope/varnishing, in the Varnishing section of Chapter 5. If you've followed the directions, your U-2 should look more or less like the prototype (Fig. 7-27). Note that the prototype was also fitted with experimental wing spoilers (Fig. 7-28).

Flying the Mitchell Super Wing U-2

8

The Mitchell Super Wing U-2 utilizes a Mitchell modified Wortman Laminar flow wing. This 34-foot wingspan airfoil has an area of 134 sq. ft. It's enhanced with the two outer panels designed with 6-degrees dihedral, and the spars swept back 12 degrees (Fig. 8-1).

Like its B-10 stablemate, the U-2's wing is outfitted with stabilators that are 8-foot long, with 12-inch chords. The washout that's normally required to prevent any flying wing from tucking is accomplished by setting these separate airfoil stabilators at a negative setting of 2-36 degrees, with no positive setting at all. They are modified from the norm by being placed upside down, attached to, but separated from, the trailing edge of the main wing. Therefore; what is normally thought of as positive is actually negative.

What this amounts to is an extremely strong wing, rated at plus or minus 10 G's; a wing that has to be rotated to the unusually high pitch attitude of 22˚½ degrees before it burbles and stalls. A wing that encloses the pilot within the wing itself, has more than a 26:1 glide ratio, optional retractable gear, and one that can handle different engine modifications.

Further modified with wingtip mounted rudders, the Super Wing gives excellent turn coordination. These rudders can either be used separately for turning, or they can be applied together as air- or speed-brakes or for drastically quick descents.

Fig. 8-1. U-2—low profile.

Assembly and Teardown

With the exception of detrailering and trailering, the U-2 goes together exactly like the B-10. Because of the Wing mods for pilot enclosure, it's near impossible to cartop the unit. In its stead, a trailer is used. However; it remains a one-person operation.

Detrailering is as simple as loosening the ship, dropping the tail gate and rolling the Mitchell off the ramp (Figs. 8-2 through 8-5). Trailering is just the reverse.

Once on the turf, take out the six Quick-Pins (Q-Pins) holding the folded wing panels in position (Fig. 8-6). Then reinsert the pins to keep the unfolded wing panels locked in line—one on each side for the leading edge, upper and lower spars—and attach the wingtip rudders with two more Q-pins, one for each rudder unit. Teardown at the end of the flying session is a reverse of the above and just as simple.

Walkaround Inspection

The first step prior to any flight is a "preflight" checkout, which includes a walkaround inspection. On the Super Wing, begin this inspection at the cockpit.

Remove the bubble canopy for access. Make sure the ignition switch is in the "off" position, and that the engine compression release is pulled—just pop it forward a couple of times. Check for control stick movement by cycling it forward and backwards and from side-to-side. At the same time, affirm the stabilators on the trailing edge of the wing respond accordingly. Look over all control linkages, including those for the stick, rudder pedals and brake unit. When inspecting the brake, ascertain there's no binding when the plunger is tested, and that the brake comes in contact with the wheel.

Fig. 8-2. U-2 rides sideways on its trailer.

Now, out of the cockpit, begin a normal walkaround to the right, inspecting the right wing unit. With any wood and fabric aircraft, look over the leading edge and surface area for holes, rips or tears, etc. Where the wings fold/join together, you have three locking pins—confirm they are in place, safety-wired or taped in position. Walk slowly on out to the wingtip rudders. Any ribs that have been broken in the wing will be readily identified through the fabric.

Open the wingtip rudders, checking for movement ease and, also, the piano hinges connecting the actual rudder to vertical stabilizer. Assure the rudder secure pin is in place and taped over. Insure rudder cable attach point and pin are in place and working.

Moving to the trailing edge of the wing, the stabilator comes first. Check its overall appearance, much as you would the wing. Then establish that tube linkages and attach points are safety wired. Now, move on to the engine.

Engine inspection is very easy, as everything is exposed—there's no cowling to remove or hatches to open. Check the engine

Fig. 8-3. U-2 unloaded and ready for assembly.

Fig. 8-4. U-2 wing panels unfold easily.

mounts, and confirm that the attach bolts are tight, with little or no play. Check the throttle, choke and compression release cables. The compression release is the only cable that will have any play in it. Examine the fuel line, making sure fuel is traveling from the main tank to the carburetor. Insure all belts are tight on the actual drive/reduction system. Inspect the propeller, as normal, and affirm prop attach assembly bolts are in place and secure, and that there are no cracks in the attach hub itself. Inspect the wire/float fuel gauge for proper operation and reading.

Fig. 8-5. U-2 with wing panels unfolded.

Model #
143-204 202

Serial
0162 10600

Progress on around the left wing's trailing edge, working around to the front leading edge, and back to the cockpit area, as you did with the right wing, except in reverse order.

The walkaround is, in reality, a simple affair, most of it being completed while you're putting the U-2 together for the flight. It's good idea to have a walkaround check sheet. In this manner, you can check off the items as they're inspected, either while putting the Mitchell together or once assembled.

Startup

Engine fire-up can be accomplished with a single body—pilot only—but, it's easier if you have an accomplice to assist. Unless otherwise stated, assume a pilot-only operation.

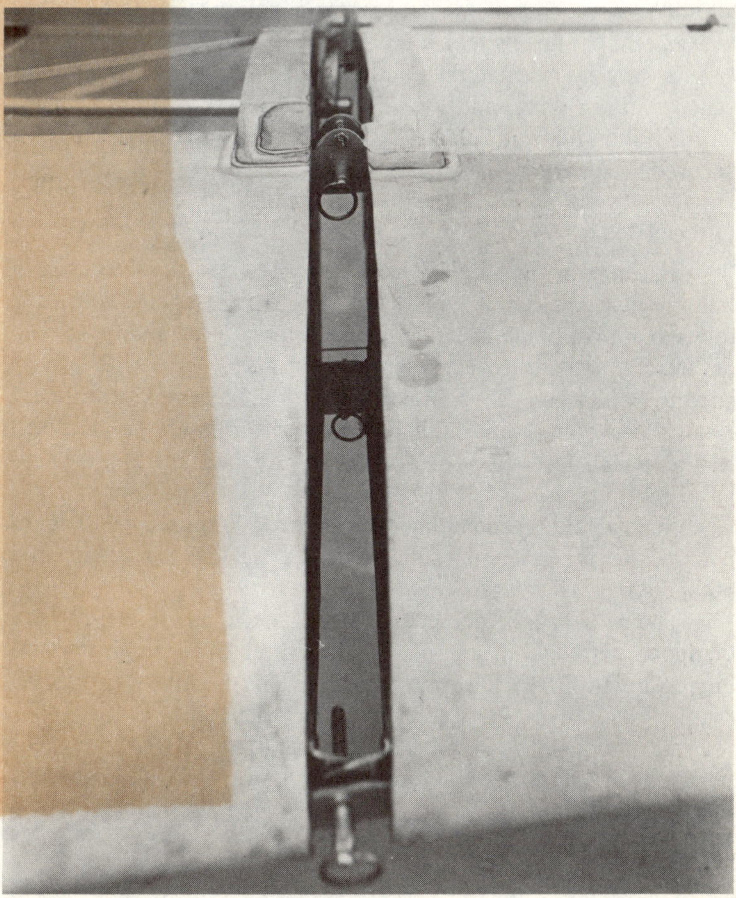

Fig. 8-6. Quick pins.

Enter the cockpit and belt in. Pop the compression release a couple of times—otherwise it's sure hard pulling the starter rope through. Reach over your left shoulder with your left hand and engage the starter pulley. Use the right hand as an assist by holding it over the left hand's clinched fingers.

Now, with the choke full "on," and the switch "off," pull the engine through 3-4 times. Once that's been accomplished, return the choke to the "off" position, turn the ignition switch to "on," and crack the throttle half open. The lever-type throttle is located inside the "elbow" compartment. The engine should fire on the first pull-through.

When the engine starts, rpm will jump to 4000-5000. After it catches and smooths out, bring the throttle back to 3500 rpm, and let it idle there. Also, as soon as it starts, release the starter rope, and depress the nosewheel brake.

Once you've accomplished the above, reach over, pick up the canopy off the wing, and lower it in place. Make sure the front attach bolts slide into place and that both side latches are fully secured. You're now ready to taxi.

Taxi and Runup

Taxiing the Mitchell Super Wing U-2 is like taxiing any tricycle-geared airplane with nosewheel-type steering. There are few exceptions, one of them being the lack of ability to use differential braking to assist in sharper turns. The U-2 is only supplied with a nosewheel brake. You'll find, however, that normal rudder pressure is more than sufficient for guidance, and the rudder pedal/nosewheel link-up is very responsive to toe pressure.

Normal taxi is accomplished at about 3500 rpm, depending, of course, on terrain. For soft or rough fields, you might find 4000-5000 rpm necessary.

The actual runup doesn't consist of much, as the engine isn't equipped with dual magnetos. Holding the brake, simply run the engine up to 6000 rpm momentarily. This is where the tuned muffler takes effect. Then return to about 4500 rpm.

Again check the stabilators and wingtip rudders for maneuverability. Monitor the wind, make sure there's no traffic, and taxi onto the runway.

Takeoff, Climb and Cruise

Once you've lined up on runway centerline, advance the throttle to 9000 rpm. Upon first applying full power, expect only

8500-8700 rpm. However; rpm will creep on up to 9000 as the U-2 accelerates.

When speed builds to 25 mph, rotate the nose to a slight upward angle. Actual liftoff occurs right at 35 mph. Hold the craft in ground effect and let speed build (Fig. 8-7).

Normal climbout is accomplished at 45 mph, with the engine turning 9000 rpm. This setting delivers a minimum of 350 fpm climbout (Fig. 8-8). During the rise to altitude, keep a close watch on engine temperature. Just a few hundred rpm can make the difference between max efficiency and melting the powerplant altogether.

At altitude, still holding 9000 rpm, let the U-2 nose over to level flight and accelerate. This rpm will net approximately 75 mph. Bringing it back to 8000 will give you a clean 65.

The only consideration here is airspeed and engine temperature. Continuing at 9000 rpm, and using a 44″×26″ pitch prop, you're going to heat the engine up a little. Bring the engine back to 8000 rpm, and it'll run 200-225 degrees F—it'll melt at 400 degrees F. During the climb, holding 9000 rpm, the engine will be heating to 325-350 degrees.

As previously stated, at 9000 rpm, in level flight, airspeed is about 75 mph, and at 8000, airspeed is reading 65. If that's too fast for you, and you want to slow down, bring the power back to 5500-6000 rpm, which will net you around 40 mph indicated.

You'll find that there's no trim on the Mitchell stock version, and during cruise flight a constant back pressure of 4-5 pounds is required. A release of this back pressure will drop the nose, and if not caught, will go to the vertical. A number of builders have positioned foam padding or bungee chords to hold the stick in

Fig. 8-7. Ground skimming.

Fig. 8-8. U-2 in flight.

neutral for cruise. Others have invented their own trim tab designs to relieve the back pressure. With this type setup, if you want to go down, you simply apply forward pressure. When, and if, you let go of the controls, they should return to the "neutral" position, keeping the ship from diving. Of course, with this method, you'll still have some amount of back pressure for climbs.

If, for some reason, you're in level flight and lose power, the nose will pitch up rather drastically. This is because of the thrust-line of the pusher propeller. When operating, it actually pushes the nose of the craft to the neutral or level position. Lose this power, and the craft tries to assume a non-powered flight path, which is up. The reverse of this is also true, so expect the craft's nose to go down or lower when power is added.

In level flight, you can, however, maintain it there, hands off, by simply pressing your knees together, clamping the stick between them. With the low-pressure force of stick activation, this mode can be maintained for long periods.

Rolls, Stalls and Spins

Stabilator roll in level flight is a little slow. But with the correct amount of rudder thrown in, the U-2 is quite responsive. It's more affected by the rudders than stabilators.

If the airplane should stall, the left wing will always break first, and the machine will roll in that direction. Recovery is only a matter of relieving a small amount of back pressure—you don't have to shove it forward—and the use of opposite rudder. Even with a full-blown, nose-through-the-horizon stall, recovery is attained easily within 150 feet, 200 tops.

Spins are corrected in the same easy fashion as stalls. When the wing breaks to the left, and you let it get established, it'll tip the bird over, with its nose going straight down. It can be described as similar to the beginnings of a "split S."

Opposite or right rudder will stop the spin immediately. I say "right" only in the sense that the craft will stall or spin to the left only in almost all situations. Therefore, corrective action will be the right rudder, right stick, and release back pressure just a mite. When I say "just a mite," I'm only talking about ½" at most or forward movement.

Any abrupt and drastic forward pressure on the stick will only aggravate the situation by throwing you forward, possibly into the canopy. So don't simply let go of the command stick, but guide it forward only the amount it takes. Remember, opposite rudder will get you out of the spin. What you are doing with the stick is breaking the accompanying stall. Then it's just a matter of recovery, and waiting for your heart to stop pounding in your throat.

With both stalls and spins, there's plenty of warning of their encroachment in the form of buffeting. The closer it is to those conditions, the more aggravated the buffeting. Also, the craft must be in an extremely nose-high attitude to enter a stall or follow-on spin. In fact, it'll be so high, you'll feel as if the machine is going to slide backwards or topple over onto its back.

Fig. 8-9. U-2 can be used as a sailplane, as well as a motorized airplane.

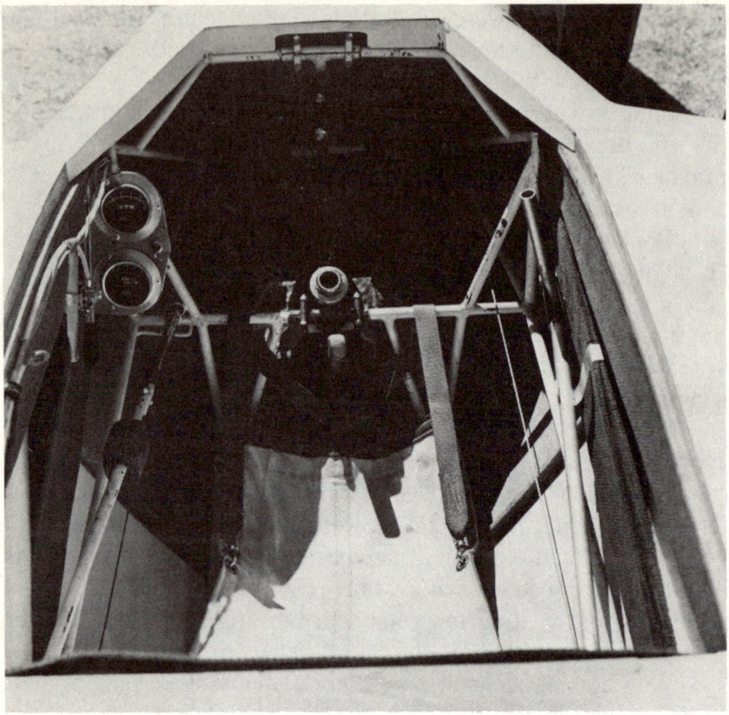

Fig. 8-10. Inside the U-2's cockpit.

Fuel, Range and Inflight Starts

The common stock fuel tank holds about two gallons of go-juice, with only seven quarts usable. That's enough for about 45-minutes of flight time. That is, of course, running it at 8500 rpm or 65-70 mph. Using those figures, you come up with a safe range of 50 miles, with some fuel/time left. However, using the bird to catch thermals, upslope convection and cloud inflow, its range—both time and distance—is up to you (Fig. 8-9).

To restart in the air, simply use the same technique as when on the ground, with one exception. Starting, either on the ground or when airborne, requires the use of both hands. Use the command stick/knee method described earlier. You need to hold the craft in a slightly nose-low position, because of the shift in CG when reaching back and pulling the starter rope through.

The CG is only a couple of inches wide. Once you're established in the seat, you can't slide forward or backward without drastically affecting it.

Descents, Approaches and Landings

Descents are usually made at about 35 mph indicated, or 55-60 may be used for high-speed let-downs. Power back to 4000 rpm for both high and low speed descents.

Make all landing approaches at 35 mph indicated. Hold the runway, reducing power with the approach, until you're running with an idle setting. In ground effect, the U-2 will fly 25 mph and below, depending on gross. That's for a stall landing. Out of ground effect, expect stall figures as stated in the stall section.

Crosswind landings are accomplished in textbook fashion—wing banked into the wind, with opposite rudder. It'll hold up to 15 mph crosswinds in the 90-degree position. Keep the crosswind configuration until touchdown.

Table 8-1. Mitchell Wing U-2 Performance and Specifications.

POWER

Engine Type: McCulloch
Number of Cylinders: 1
Horsepower: 10 8000 rpm
Propeller: Wood—Hand carved
Propeller Diameter: 44″
Propeller Type: Fixed pitch
Aviation Fuel Grade: Fuel/oil mix
Power Loading: 4.5 lbs/hp

SPECIFICATION

Wingspan: 34 ft
Wing Area: 136 sq. ft.
Length: 8 ft (9 ft/w nose cone)
Height:
Seating Capacity: 1
Maximum Gross Weight: 450 lbs
Empty Weight: 189 lbs
Usable Weight: 261 lbs
Fuel Capacity: Standard Tanks—2 gal (7 qts usable)

PERFORMANCE

Cruise: 80 mph
Stall Speed: 23 mph
Takeoff Distance: 200 ft plus
Landing Distance: 95 feet less
Best Rate of Climb: 400 fpm plus
Service Ceiling: Undetermined
Fuel Consumption: 1 gph

Landing distance is right at 250 feet, with the takeoff running as high as 400. That's on a hard-surfaced runway, of course. Soft or rough fields will take longer takeoff distances.

Braking with the single nosewheel brake is greatly enhanced by letting the wing lose its inertia, then leaning forward and grasping the brake handle and squeezing. By leaning forward, you move the CG and your weight forward, applying more pressure over the braking area. Results—more efficient braking and faster stops, using less runway.

Shutdown of the engine is made by simply running it up to 4000 rpm, hitting the compression release cable, and turning the ignition to the "off" position (Fig. 8-10). Just engaging the compression release should shut the engine down. If not, pop it a few more times, until the compression does die down.

Building the Mitchell P-38 Lightning

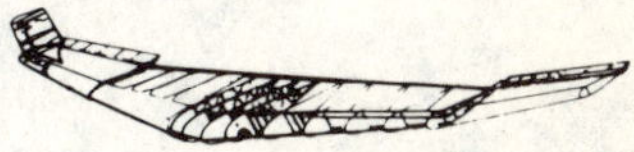

9

The Mitchell P-38 Lightning, like the Wing B-10 and Super Wing U-2, is an all-new design. All three ships were designed by Don Mitchell and were developed to fill a specific need in the homebuilt ultralight field.

The Lightning has been hailed in homebuilt circles as a pure and simple delight that can be purchased in kit-form on a Friday, put together on Saturday and flown on Sunday. It is, in fact, the answer for those wanting to construct and fly their own aircraft, but who do not want to wait from six months to five years while building.

Plans are available, being mostly full-sized, with material lists, instructions, and photos leading you step-by-step to a completed project in 110-150 man-hours. Throw in a few extra hours for personalization of the cockpit and a paint scheme, and you are on your way.

The P-38 kit is unique in that all components are prebuilt. There are less than 200 parts. It's as simple as putting together a large jig-saw puzzle (Fig. 9-1). It's from this concept that the Lightning can be assembled by four people in one day. Obviously, if there are less personnel working on the project, it's going to take longer.

The only tools required for assembling the P-38 kit are those normally associated with garage workshops, including hammer, saw, screwdriver, drill, hand-riveter and snips. Working from scratch, with plans only, you're going to need a few extra, like a hot-wire for foam forming, welding apparatus, etc., but nothing that the layman can't operate or find assistance with.

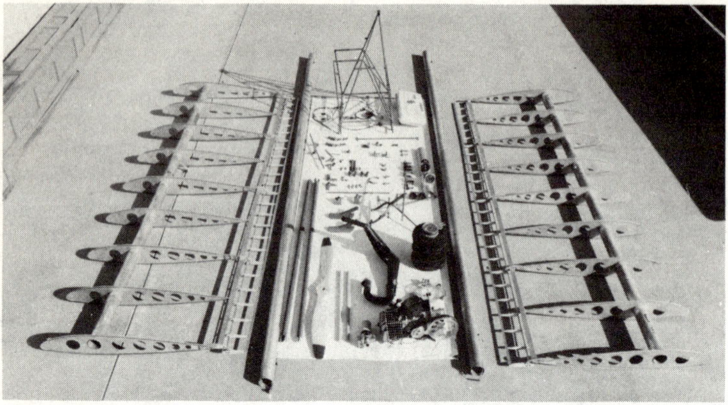

Fig. 9-1. The Mitchell P-38 Lightning kit contains approximately 200 parts which come together much like a giant jigsaw puzzle.

Fuselage

The fuselage, as such, is 16 feet long and made from two 3 ½″-diameter aluminum tubes (Fig. 9-2). Unfortunately, these tubes come in only 12′ lengths, so one of your first tasks is to properly cut, mate and splice two of these sections together, with the end product being a tube 16′ long. Two 12′ sections and two 4′ portions equal two tubes 16′ long. Take one tube (fuselage section) at a time, then duplicate the process on the other side.

These 4′ segments are "scabbed" onto the 12′ lengths by making two inner sleeves which will go inside the long and short pieces of tubing (Fig. 9-3). This is accomplished with an 18″ length of the same tubing sawed lengthwise on one side. The thickness of the saw blade cut will give the tube enough flexibility to be squeezed gently, thus narrowing the diameter so it can be roughly slid inside the main tubing halfway, and the "scabbed" 4′ length slid over the other half.

Next, drill ⅛″ holes for riveting. There should be approximately two holes per square inch. I say *approximately* because the rivets are only used to hold the material in place while the epoxy dries.

Once the rivet holes are drilled, you're ready for the mating. Clean the surfaces to be joined with MEK (methal ethel keytone), then paint the inside of the tubing and the scabbing installment with 9410 structural adhesive. Slide the sleeve in place, and hand rivet for solidity. Let bonding dry.

It's very important that these spliced components are straight on line with the main tubing section. Otherwise, the fuselage tubes

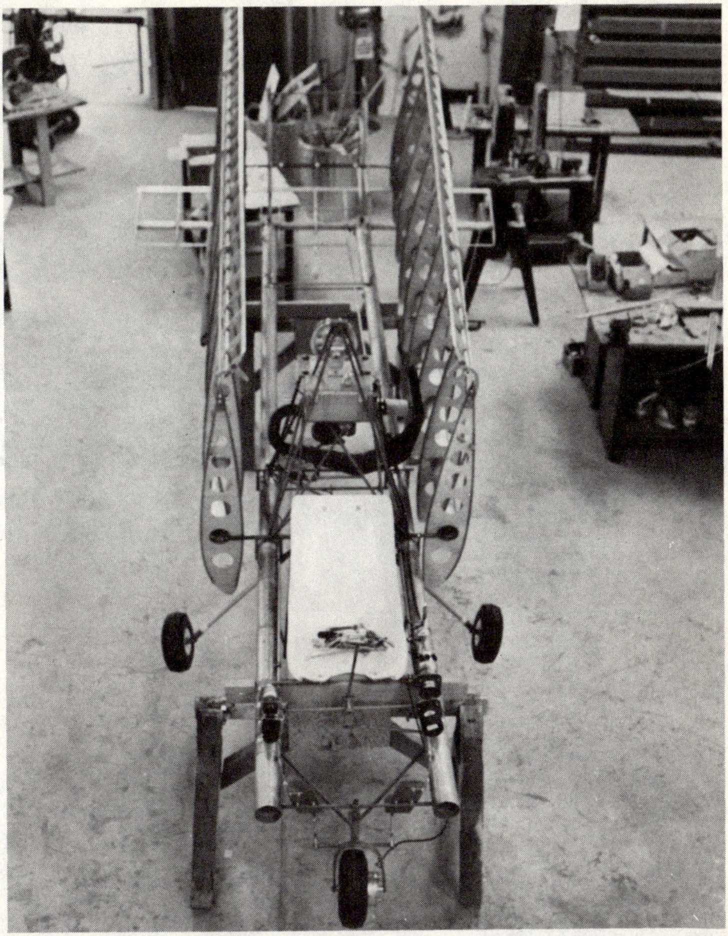

Fig. 9-2. P-38 with wings folded. Notice fuselage tubes in 2 × 4 jigs. Here final adjustments are made.

will be uneven in separation when put side-by-side for crossbracing application.

These two fuselage tubes are joined by an intricate webbing of crossbracing (Fig. 9-4). However, since welding on the actual aluminum tubes would mean "damage beyond use," a number of saddles or pads are made, to which this bracing is welded, then the saddles are riveted and epoxied to the fuselage tubing.

These saddles are made from .040 chromolly*. You're going to need 14 or so, measuring 3″ × 8″. To make these pads, cut the

*Chrome moly steel.

Fig. 9-3. Fuselage scabbed section can be seen just in front of landing gear on main tubes.

material to spec. You need an 8-10″ section of 3 ½″ steel tube to use as a jig. Then it's just a matter of heating and beating these pads to fit snuggly over the jig.

When completing the final assembly, make sure to position the fuselage tubing with the spliced sections up front. In this way, the scabbed sections will be out of sight. Also, these splices make the tubing stronger, so you might as well have it up front where it's needed.

Nosewheel Area

Start building at the front end, by making the nosewheel assembly first (Fig. 9-5). You can get the whole front end on with the use of a welding jig.

The actual nosewheel fork is nothing more than a Schwinn bicycle front end, cut and formed to size, so it will take the nosewheel and braking system. Bracing and linkages are then added to complete the structure.

To modify the bicycle fork, first heat the right side (looking nose-on), so that it's pliable. Now, straighten it to a right angle with the opposite fork. Re-bend it to take the 10″ nosewheel and brake. Once you've accomplished the bending process, you'll find the elongated fork's side shorter lengthwise than the left side. Cut the left side off to match the right.

Fig. 9-4. Nosewheel section, partially showing intricacy of webbing and crossbracing.

Fig. 9-5. Notice nosewheel fork bent outward to carry drum-type braking. Control system mixer box is seen directly behind, with military fashion control stick coming between pilot's legs.

The gooseneck mechanism above the fork has to be lengthened about 8″, so that means another piece of tubing splice-welded into position. While working with the gooseneck assembly, pull out the actual fork mechanism, so as not to damage it when heating.

This 8″ gooseneck extension is made from 1″ .049 chromolly tubing (Fig. 9-6). First, cut the existing gooseneck in half. Save the upper section with the threaded portion. Slide a wooden dowl through this neck. The dowl is used for alignment while welding. Now, weld on the 8″ extension, then replace and weld the threaded portion previously saved, and let the whole assembly cool.

Once you've got the extra 8″ of gooseneck welded in place, you need to brace it to the tubular fuselage. Nosewheel bracing and linkages are nothing more than a chromolly truss, with four support saddles for the fuselage tubes. Tubing used is ½″ × .035, with a ⅝″ crossbar.

For these attachments, you need a jig to work on. Make it from two 8′ lengths of 2″ × 4″s. They should be placed side-by-side, with 24″ centers. These 2″ × 4″s are, in essence, taking the place of the two fuselage tubes.

Place a previously made saddle on the underside most forward part of each fuselage jig, then install two more approximately 13 ½″

Fig. 9-6. Nosewheel and attending linkages. Notice the 6″ drum-type brake and its mounting to the fork assembly. Also note 12 degree forward slant of gooseneck and nosewheel.

aft of those, also on the underside. These saddles will carry both horizontal and angled crossbracing for the fuselage, as well as the nosewheel component.

A 24″ horizontal is first. Lay this tube in your welding jig. Next, tack weld the top of the gooseneck extension to its center, making sure the gooseneck is square with the crossmember. About 1″, or the threaded part, of the gooseneck should extend above the crossmember.

Now weld two tubes—one on each side—to the bottom of the gooseneck in such a fashion that they come up to the fuselage tube's forward saddles at a 45-degree angle, and they also angle rearward by 12 degrees. This latter 12 degrees allows the gooseneck and wheel assembly to angle forward 12 degrees from the horizontal crossmember.

Measure up approximately 3″ from the bottom of the gooseneck. From here, weld two more tubes—one from each side—running up at 45 degrees also, and connecting to the aft two saddles (Fig. 9-7).

You should now have one horizontally braced crossmember up front, connected to the gooseneck. From the gooseneck, four tubes running up at 45-degree angles—one on each side to the front saddles, and one on each side to the aft saddles. The bottom two tubes angle aft of the gooseneck at 12 degrees, with the upper two tubes taking whatever angle needed to reach the aft saddles.

After these components have been tack welded into approximate position, make sure all angles are correct. Then weld.

Onto the top of the nosewheel fork, you want to weld a 24″ horizontal rod for direct rudder manipulation. Six inches in from each tip, heat the tubing and bend until the tips are vertical, or form a 90-degree angle.

From the tips of each side, connect a uniball joint. Direct linkage will run from these uniballs to the rudder pedals, ending in a clevis fork arrangement. Rudder pedals are mounted to the front crossbracing, with a metal tab hanging down to take the clevis fork on each rudder pedal. These pedals are made from ¾″ pine and custom built to fit your feet.

The steering linkage is very critical and must be true in either direction to 15 degrees. If you have too much, or it's too loose, you'll be all over the runway; too tight, or too little, and you'll end up with an excessive turning radius.

At this point, the only thing you have to do with the actual brake is to weld on a 5/16″ stud coming off the inside of the right fork (looking nose-on) to hold the brake, or back of the brake plate. Then just run the brake cable back, and let it hang there, until installing cockpit instruments and controls.

Control System

The control system is, basically, two saddles on the fuselage tubes, with a crossmember 22″ long of ½″ chromolly tubing. On top of this crossmember is welded a ⅝″ chromolly tube about 2″ long. Tube openings face fore and aft (Fig. 9-8).

Fig. 9-7. Gooseneck bracing is plainly seen taking 45-degree angles to the fuselage tube's saddles— two fore and aft.

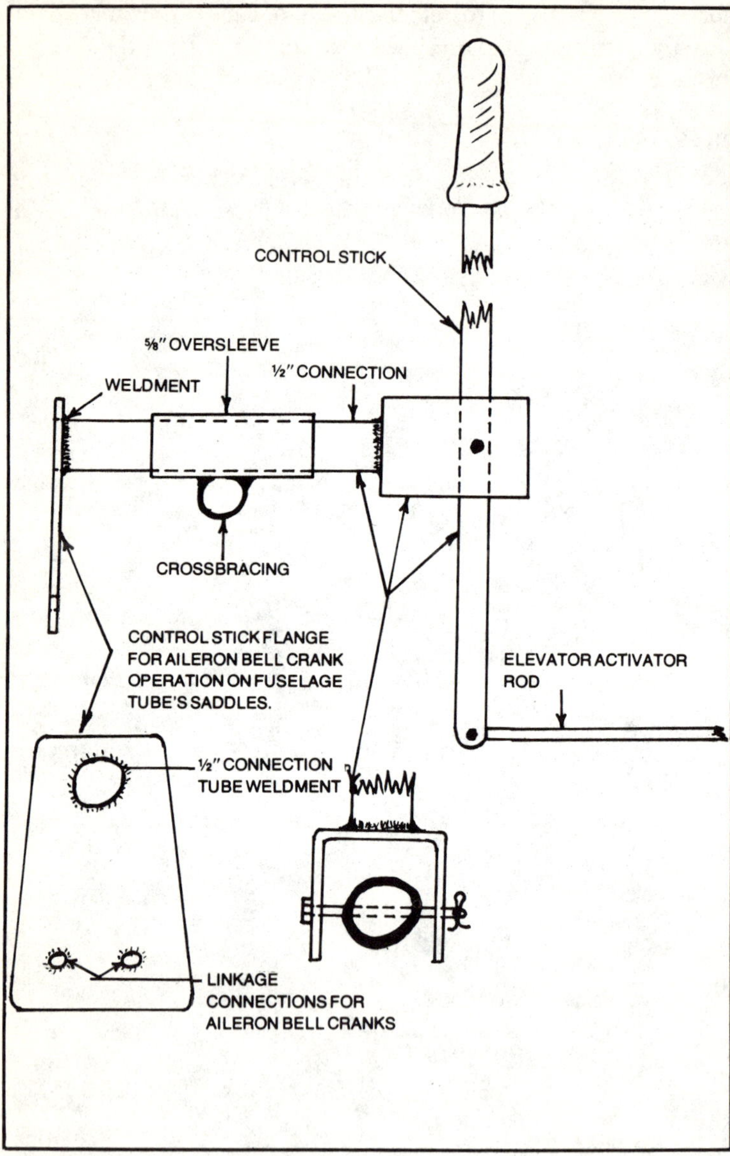

Fig. 9-8. Control system schematic.

Through the ⅝" part channels a section of ½" tubing, approximately 3" in length. Onto the forward portion of this ½" tube is welded .040 4130 chromolly 1 ½" × 3" to be bored and used as a bellcrank for the ailerons. Onto the aft end of this ½" tube is welded the U-shaped universal mechanism for the control stick. The stick is pinned vertically in the U channel so as to be rotated forward and backward, activating the elevator. Elevator linkages are rod-connected to the bottom of the stick through a clevis fork arrangement. (See detail drawing.)

On the outboard sides, underneath the fuselage tubes on the same saddles, there are studs coming down which hold the bell crank for the ailerons. These studs are ¼″ bolts cut to fit.

Once you've got the studs in there, you can rig the two interconnect rods for the ailerons, running from the outside bell cranks to the 1 ½″ × 3″ section of the control stick mechanism talked about earlier. This bell crank is very important. If it isn't aligned properly, it may interfere with operations (see plans for details).

Pylon, King Post and Attachments

The pylon and king post (Fig. 9-9) are the P-38's key components since the pilot's seat, engine, fuel tank and control systems are supported by it. All flight loads are channeled into the pylon and king post. There's a lot of detail involved, and, you might say, the aircraft is built around it.

The king post is made from ½″ .035 chrome moly wall 4130. The base of this assembly is 26″ square. The fore and aft, and the aft lateral, are made from this ½″ chromolly. The cross-section of the forward area of the base is ¾″ .058 wall.

To make a welding jig for the king post, draw out a 26″ square on a piece of plywood. Concentrate on the forward and aft crossmembers.

At the center of the forward 26″ crossmember, draw a line out vertical for 33″. This will be the height of the king post. Now, from each end of the crossmember, tack weld a tube so that it meets, and is tack welded, at the 33″ vertical center.

To the aft crossmember, you'll tack weld two more lengths of tubing—one on each corner. However, these will be a different length than those forming the forward triangle. The reason for this difference is that the forward triangle is vertical, and the aft triangle is angled forward to meet it. So prop the forward triangle vertical, then measure the aft tubing to lean forward to a common point. Then tack weld the aft triangle.

With the fore and aft tubing triangle tack welded, prop them in their approximate positions, and tack weld the left- and right-hand side tubes into position, connecting the two triangles at the base.

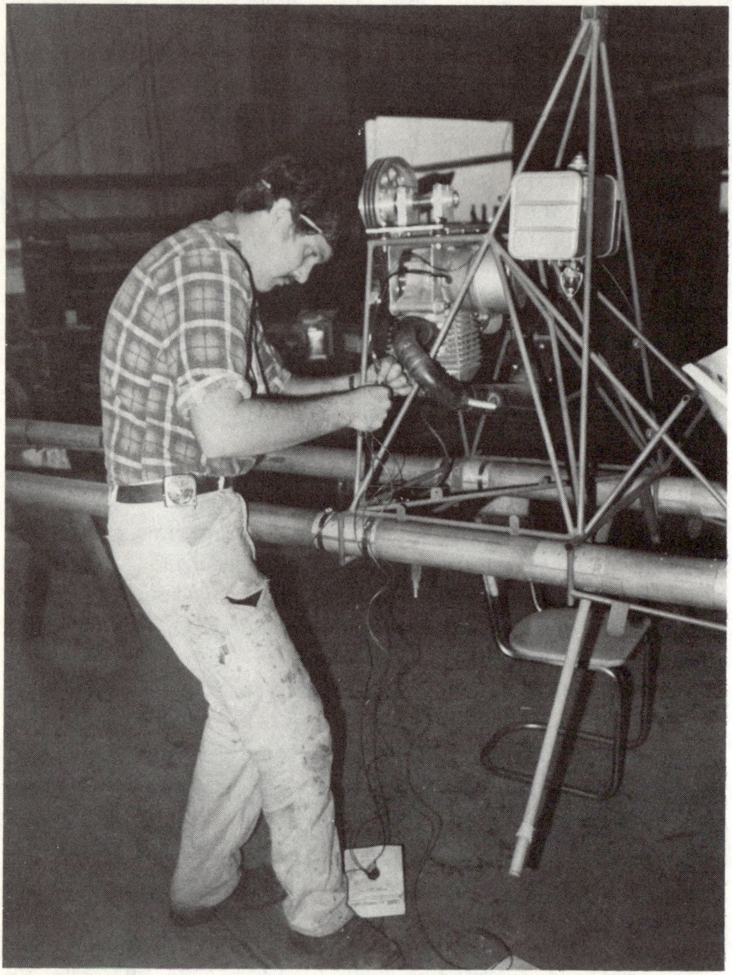

Fig. 9-9. Pylon/king post, with 250cc Honda engine mounted. Notice aft triangle leaning forward and joining forward triangle.

After you've rechecked the angles and lengths, weld for permanent bonding.

Where these two triangles come together, on top, build and install a 1″ square chromolly tube. Weld this square to the triangles.

At the base of each bottom corner, weld a square, using ½″ chromolly (Fig. 9-10). This square will be approximately 4 ½″ inside-to-inside. They are connected to the king post base corners facing fore and aft. The purpose of these squares is to hold the main fuselage tubes—the tubes passing through the 4½″ squares. There are few loads on these tubes at these points, however, upper and lower saddles are applied where each corner of the king post, or the 4½″ squares, rest.

About two-thirds of the way up vertically, or 22″, level with the line of flight, tack weld two—left and right—⅝″ .058 chromolly tubes about 20″ long. They attach at the 22″ vertical point on the aft triangle, and extend aft. If you looked at the craft from a top angle, you'd see these two tubes running rearward, with about 10″ between them. This space is where the engine is to be mounted. Two vertical tubes are also tacked from the left and right aft bases to the aft end of these rearward-pointed tubes, leaving the aft horizontal area open for the engine mount itself.

To these two tubes are connected the four engine mounts. They are bolted on, rather than welded. Use AN4-26 stock and Honda Lordmounts. The engine is then mounted to the plate that joins the two tubes aft (see plans for details).

Where these two tubes meet the aft triangle, run a crossmember horizontally across and tack weld. Now run tubes from the center of this crossmember to the fore, left and right base corners. Also, connect tubes running from the left and right forward base corners to the intersection of the tubes at the 22″ point on the aft triangle.

Fig. 9-10. Rear base saddles, with 4½″ squares, through which the fuselage tubes go. Notice elevator bellcrank system as crossbracing between in this mount beef-up.

At the 22″ mark—left and right—on the aft triangle, run tubes through the forward triangle at approximately 45 degrees (Fig. 9-11). Where these tubes join the main fuselage tubes, put saddles. This left and right tubing is the back support for the seat. Where the forward portion of the seat rests on the fuselage tubes, you also want to install a crossmember just under its lip for support.

Main Landing Gear

For the first 7-8′ of the fuselage, get everything worked out on a jig, just as it's going to be on the airplane. When you slide the prebuilt components onto the two fuselage tubes, you want to do it once and *only* once, because aluminum is very susceptible to damage—scratching, kinking, wrinkling, etc.

The main landing gear supports are nothing more than 3′ lengths of ⅞″-diameter fiberglass rod. This gives the undercarriage both strength and durability. The whole assembly has been drop-tested from 15′.

The main landing gear sockets which take these ⅞″ fiberglass rods are 1″ × .049. They are welded onto the king post assembly at the corner bases of the forward triangle and flay outward at a 45-degree angle. Into these sockets, the fiberglass landing gear rods will later be glued.

Attaching the main wheels is one of the last things to complete, as the full weight of the aircraft, including the pilot weight and that of fuel/oil mix, must be taken into account and adjusted for.

When you come back to putting on these wheels, first slip them on and hold in position with masking tape (Fig. 9-12). Either load up the structure with sandbags, or put a person in the cockpit to simulate landing and taxi loads.

Now, simply roll the craft forward and backward until you get the wheels to do what you want—so that the glass rods splay out like they would on landing. Adjust them for toe-in or toe-out.

Once you've got this angle set, drill a ¼″ hole, according to the torsion brace. This brace runs from the wheel connection to a dampener set-up at the corner of a crossmember at the bottom of the seat. Then it's just a matter of applying 9410 adhesive and allowing to dry.

Rudder

From the king post aft, all you've got is the two fuselage tubes running back, until you get to the tail feathers (Fig. 9-13). This

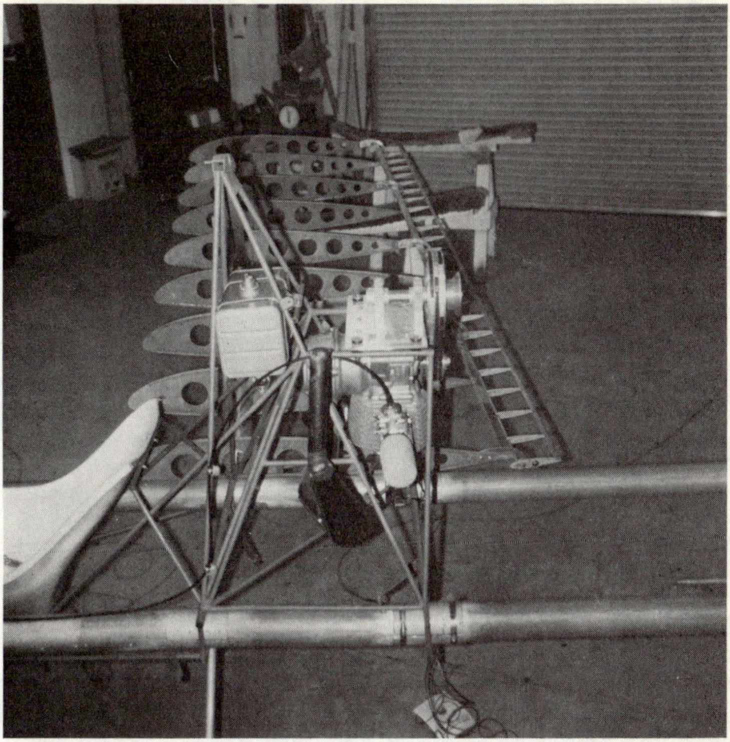

Fig. 9-11. Notice layout of tubing structure and forward sweep of the king post and their joining at a common 1″ square tube weldment. Also, the mount reinforcement of the main tubing.

section is a real homebuilder's delight. The rudder and vertical fin will be made on a jig, prior to installation.

First off, there are 24 small socket fittings that come together to make the rudder. Some of these sockets will be 1″ × .049, others ½″ × .028, or a mixture of the two, so make them as needed. Then it's simply cleaning the aluminum with sandpaper, bonding and riveting. You're going to make each component in sets, including the sockets.

Draw out in pencil on a piece of plywood two vertical 35″ lines, about 3″ apart. From the bottom of each line, draw another 90-degrees horizontal. Extend this line out quite a ways. From the top of the 35″ line, draw another line coming down at a 45-degree angle, until it crosses the horizontal. Now come up the vertical line halfway, or 17 ½″, and draw a horizontal line across, until it intersects the 45-degree line.

Fig. 9-12. One wing folded and one unfolded. Wheels are attached to ⅞″ fiberglass rods with masking tape for final tracking adjustment prior to bonding.

Except for the vertical post, which is ⅞″ aluminum tubing, the vertical fin members are of 7/16″ aluminum. Unless otherwise stated, all sockets will be .028 chromolly tubing. Now weld up sets of sockets per plans, or to form.

As the length for the complete rudder assembly, including the fin, use approximately ⅓ the distance of the fuselage tube. This

Fig. 9-13. Twin rudders and elevator. Notice post sticking out of rudder's vertical fin for compression strut/wing tiedown when streamlining for trailering. Also note bellcrank system for both rudders and elevator. Tubing angles —15 and 45 degrees — can be seen through fabric of the rudder and vertical fin.

will be about 6′ from the forward edge of the vertical fin to the trailing edge of the rudder.

As with the vertical fin, draw out the actual rudder assembly on plywood, then cut and match with component parts. Draw two vertical lines 35″ in length and, again, about 3″ apart.

From a point 3″ up from the bottom of this line, draw an angle up at 15 degrees for approximately 24″ (Fig. 9-14). From the top point of the vertical line draw a line horizontally aft, or 90 degrees, for about 12″. From the tip of the 24″ 15-degree line, draw one

Fig. 9-14. Tubing and ribs of the horizontal stabilator (elevator) and vertical stabilizers. Tail skids were later modified. Bottom of moving rudder begins 3″ up from end of fuselage tubing and slants up 15 degrees to give elevator moving room.

vertical, until it passes the 12″ horizontal line at the top, if it were extended out.

Mark a point 16″ down from the top of the vertical line. From this point, draw a line angling up 15 degrees aft, until it intersects the aft vertical 32″ line. Mark another point, about 6″ up from where this mid-horizontal line intersects. Now freehand a semicircle from this point to intersect the end of the upper 12″ line.

Now with the working portion of the rudder assembly drawn out, simply fill in with the properly cut tubing and sockets. The forward vertical is ⅞″ aluminum tubing, with all others 7/16″.

When making the curved portion of the rudder, just bend into position as close as possible. Let the tubing carry its own curve without crinkling or kinking.

You'll notice from the drawing on the jig, and while assembling the working rudder, that it actually starts 3″ *above* the base. This space is necessary to carry rudder activator rod connecting the two vertical rudders (see plans for details).

This rod is made from 7/16″ aluminum, and is attached horizontally to the vertical tube with a clevis fork arrangement. Hinge-fittings are 1″ × .049, with 5/16″ .065 chromolly as hinge pins. The moving rudders are attached to the aft vertical of the fins.

The vertical fins are spaced parallel at 23 ½-24″, or along the centerline of the fuselage tubes, with saddles fore and aft (Fig. 9-15). They are connected with a forward base crossmember of 7/16″ aluminum, socket-bonded and riveted. The same type crossmember is installed at the aft base, aft mid-vertical and aft top of vertical fin. Also, an angled crossbrace is positioned from the aft base corner to the opposite aft mid-vertical point.

Through a system of bellcranks, the rudders operate with cables running through the fuselage tubing, directly to the rudder pedals forward.

Wings

The wings, elevator and ailerons are made very similar to each other (Fig. 9-16). The whole theory is to build the ribs, and slide them into place on their prospective-sized spar tubing.

Wing ribs are cut according to plan from 1/16″ (1 ½mm) birch plywood, with the grain vertical. There are a total of 18 ribs—all exactly the same size. Cut them all at the same time. Wing chord is 48″ or 4′, with wing depth being 15 percent of the chord, or approximately 7 ¼″.

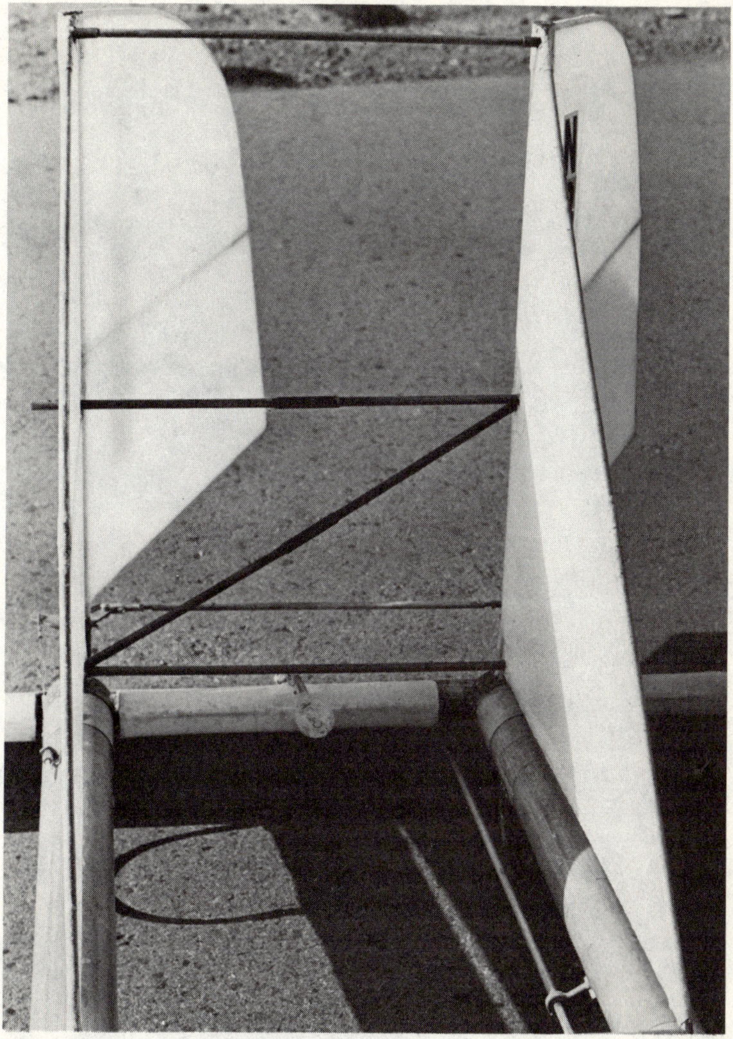

Fig. 9-15. Twin vertical rudders and bracing. Elevator is activated with push/pull rod running along fuselage tube, while rudders operate from cables channeled through the tubing itself. Notice crossbracing and sockets.

Once the ribs have been cut to size, concentrate on the wing tube hole and the lightening holes. The tubing hole is 3 ½″ in diameter and the lightening holes vary from 3″ to 1″.

The tubing hole is located 25 percent back on the chord line of the rib. This equates to 12″ to center, with the vertical center

Fig. 9-16. The wing's 5-degree forward sweep is very obvious here. Notice wing spar and lightening holes imprinted against wingtip fabric.

being the chord line. After cutting the 3 ½″ tubing holes, sand or file so that they fit snuggly, or slide roughly onto the tube (Fig. 9-17).

If these holes are too tight, the ribs deform as you slide them on; if too loose, you simply can't glue them in place. Therefore; sanding is better than filing.

Fig. 9-17. Overall view of wing ribs, lightening holes and aileron hung on trailing edge. The sawhorse under wingtip and compression strut connected to wing's inboard section help in fixing the wing's 6-degree upward dihedral.

If you look at a finished rib from the side, you will see the tubing and lightening holes in this order along the chordline—3″ hole, 3½″ tubing hole, 3″ hole, 3″ hole, 3″ hole, 2″ hole and a 1″ hole. Drill these lightening holes equally-spaced.

Rib capstripping is special, being ¼″ × 5/16″, with a groove cut right down the ¼″ side center (Fig. 9-18). This groove is obtained from a very fine plywood blade. Capstrips run from the trailing edge of each rib to a point 4″ before the leading edge curve.

The groove in the capstrip is for gluing the capstrip to the rib. Pour adhesive down the groove, and clamp the capstrips to ribs with strong rubber bands, and let dry. It should take about three minutes per rib.

Once you've reached the stage of having all of the ribs and capstrips bonded and holed out, clean the wing tubing. This is done by sanding with very fine sandpaper. Be sure to sand *lengthwise* on the tube and not *around* it. Next, clean the tubing with MEK, and slide the ribs into their approximate positions on the tube.

You'll want to assemble both the right and left wing at the same time, for reasons soon to become apparent. Keep the wings and components separate, but visible. In other words, don't make one wing, and stick it out of sight while you're building the other.

Follow the plans for rib spacing. Measure rib spacing on the tube and mark it in pencil. All marks are rib centers, with the exception of the wing root rib. From the wingtip rib, spacing will be 18″ centers, with the exception of the last rib, which will be 17 ½″.

Fig. 9-18. When applying wing rib capstrip, uniformity of length is more important than actual length of capstrip from rib leading edge.

Fig. 9-19. Wing ribs, with spacing blocks to hold ribs firm while epoxy 9410 structural adhesive is drying. These blocks can be made from any type material, as long as lengths are correct.

After you have the ribs in place, cut a bunch of blocks from scrap to fit between, and hold the ribs securely in place (Fig. 9-19). Slide the ribs to one side or the other of their position marks about ¼″, and clean again with MEK prior to bonding.

These blocks are needed because, as you slide the ribs onto the tube, they tend to roll this way and that, or walk around a bit. They hold the ribs stationary. Now it's just a matter of painting with adhesive, fit a rib, and place a block; paint a rib, place a rib, slide in a block; etc. All of this is accomplished on the workbench or floor, before attaching the wings to the fuselage.

Another reason for completing this section on the ground is that each wing's outboard tip is washed out by some three degrees. Therefore; the outboard ribs take a special tape-board that fits under their trailing edges. With both wings on the floor at the same time, you can look at one and say, "That's the left wing," or whatever to keep them separate.

In essence, then, what you're doing is sliding the ribs in place, blocking them, and lifting the trailing edge—making sure everything is *square*. Now, simply walk away and leave them alone for a while.

Ailerons

The ailerons come together much like the wings and elevator (Fig. 9-20). All ribs are cut according to plan. They are 6″ long and ⅞″ high. However, unlike the main wing and elevator ribs, they have no cambered leading edge. Instead, the leading edge is a ⅞″ chromolly tube, 12′ long, which carries the leading edge curve.

There are 25 such ribs, with 5 doublers, making a total of 30. Cut these ribs and doublers 7″ long, for reasons soon to be apparent. As with the wing and elevator, cut all ribs at one sitting.

Since there are no lightening holes to be drilled, it's just a matter of drilling the ⅞″ hole to carry the ⅞″ leading edge tube. Drill this hole so that the forward radius ends up at the 6″ mark. Ribs and doublers are cut tapered up from the rear to a 1″ thickness. After bonding, they are sanded down to the ⅞″ thickness of the tube.

Spacing on the ribs is 6″ from center-to-center. Clean the ⅞″ tube with MEK, then mark off 23 of these 6″ rib placements. Inside every 6th rib, beginning from the end ribs, pencil a mark for a doubler. There is approximately 1″ between the rib and doubler, center-to-center. These doublers are for wing attachment fingers. Slide the ribs and doublers into position over the witness marks, and check for uniformity. After you're satisfied, move them to one side, and clean with MEK. Apply 9410 structural adhesive for bonding, and let dry.

When you've applied the glue and aligned the rib, drive nails on each side of the rib. The nails should be implanted on both sides of the rib in such a manner as to hold it in alignment until the epoxy sets up.

After the glue has dried, cut the rib and doubler forward section back to the ⅞″ tube's centerline. In effect, this means the tube will be the aileron's leading edge. Now, sand down the upper and lower portions of each rib, until it angles uniformly from trailing edge to, and disappears at, the ⅞″ tube's centerline.

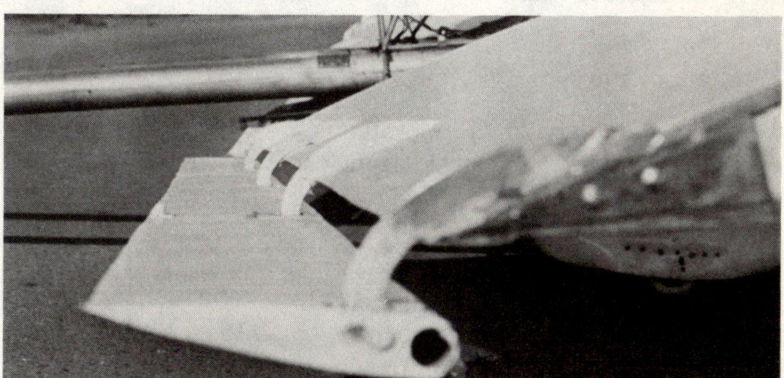

Fig. 9-20. Ailerons are hung behind and below the wings' trailing edge. They fly in their own airstream, making them extremely effective at slow airspeeds.

Prior to skinning the ailerons, cut out properly curved fingerlings to go from the wing to the doublers, according to plan. There are five of these hinge fingers, cut from plan templates. The aft section of wing ribs will have to be beefed up to carry the load. These fingerlings attach to every other wing rib.

Either use a right-angle drill to cut holes in the doubler and rib on actual assembly, or leave the two parallel ribs loose—to be drilled and glued later. Use AN3-12 bolts.

An activator rod comes off the inboard end of the ailerons and, through a clevis fork arrangement, runs to the forward control system. This rod is made from 7/16″ aluminum.

Once you've got the hinges attached to the ailerons, the ribs bonded, you want to skin the whole affair, with the exception of the leading edge tube. Use .8mm ply. The ailerons are still flat on your workbench or plywood.

As you apply the skin and bond, staple the .8mm ply along rib centerline to insure tight bonding. Work the skin forward from the trailing edge to where you have approximately ¾″ sticking beyond the leading edge tube.

Apply epoxy under the leading edge portion of the skin—where it sticks out beyond the tubing. From the top, drive small-gauge finishing nails through, and into, the workbench or plywood. This makes the skin take a slight curve down and around the tube.

Let this dry in position. Afterwards, cut off the excess ply forward, where it comes away from the tube, and sand flush for even curvature. The completed ailerons will be attached to the wing and adjusted later.

Elevator

Like the ailerons, the elevator is a symmetrical airfoil. In other words, the upper camber is the same as the lower camber. Like the wing ribs, however, those of the elevator take a tubed spar and lightening holes.

The elevator is 8′-long, 18″ wide and 2 ⅛″ deep. It takes in 9 ribs and 2 doublers, so cut 11 ribs. The main tube running through the ribs is 2″ diameter aluminum. A 7/16″ aluminum tube completes the elevator's leading edge, much like the ailerons.

As with the wing, come back 25 percent of the chordline to find the center of the 2″ tube hole. Drill these out right on the mark, then sand gently with light sandpaper until the ribs will slide

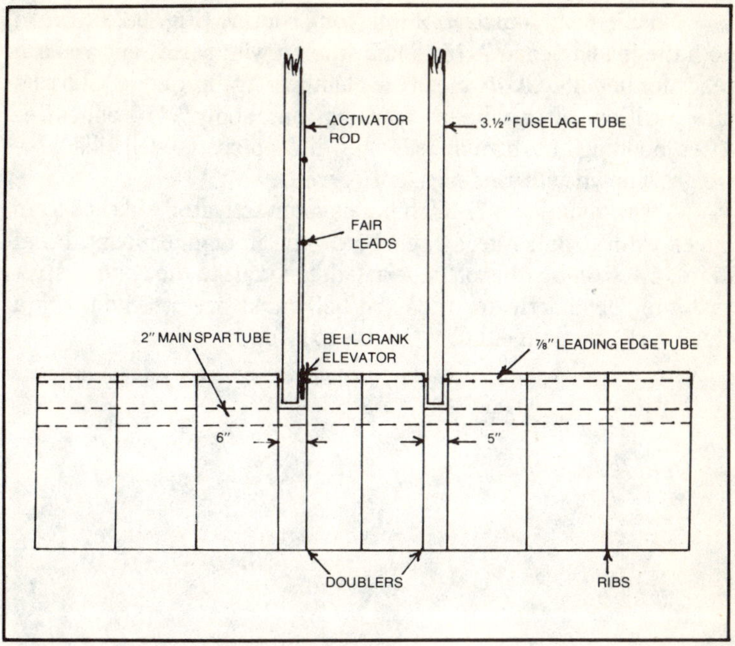

Fig. 9-21. Horizontal stabilator schematic.

roughly onto the tube. Remember, you don't want too much slippage.

Also on the chordline, approximately ½" back from the leading edge of each rib, drill the hole for the 7/16" tube. Again, sand each hole, until the rib slides roughly over it. Keep to the chordline, and drill three 1" diameter lightening holes between the 2" tube and the trailing edge. These lightening holes are evenly spaced.

Moving inward from both elevator tips, a doubler is placed inside the 4th rib from each side (Fig. 9-21). These doublers are spaced 5" inside the 4th rib from one side and 6" inside the opposite 4th rib. This room between rib and doubler makes space for the fuselage tube attachments. The 6" space is wider to take in the elevator activator rod. Therefore, the leading edge 7/16" tube must be stopped from the 4th rib from each side to the doubler on each side, then continued from the doublers across the center rib.

Across the trailing edge of the elevator is bent to form aluminum sheeting. Cut the sheeting, and bend it to carry the ribs' trailing edges. This sheeting comes forward, both above and below the ribs, for 2".

Now it's just a matter of following routine (Fig. 9-22). Clean both the leading edge 7/16″ and 2″ tubing with MEK, and you are ready for bonding. Also, clean the aluminum trailing edge. Slip ribs into position, after witness marking, and apply 9410 adhesive. After bonding the trailing edge—while epoxy is still pliable—weight it down with sandbags, until cured.

A 9′ 6″ aluminum 7/16″ tube elevator activator rod runs from the elevator to bellcrank system below the aft engine mount. From there, a 4′ 6″ tube runs on underneath the craft to the control stick system. These activator rods and bellcranks are attached with a clevis fork arrangement. Cut lengths to form.

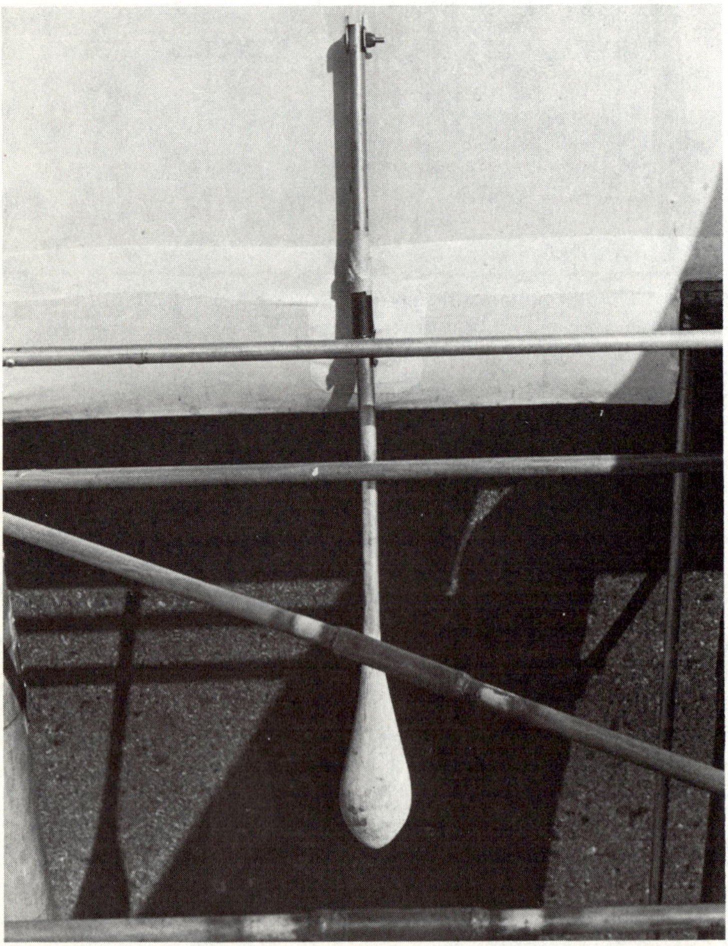

Fig. 9-22. Elevator (horizontal stabilator) balancing weight.

Fig. 9-23. Overall view of the P-38 ready for covering. Notice it's all resting on the two sawhorses. This is how it's built. The two tubes go on the sawhorses, then, working from the nose rearward, until complete.

Assembling

Building it is one thing—putting it together is another. To begin with, it's easier if you first cut 3 ½″-diameter half-holes in a couple of 2″ × 4″s, in which to rest the fuselage tubes during assembly. Place these 2″ × 4″s on opposite ends of your workbench or saw horses, and insert the main tubes (Fig. 9-23).

Start by sliding everything onto these two tubes—engine mount, king post assembly, etc., making sure you don't scratch those two tubes. You can leave the control system off for now, as it may get in the way. Your primary objective at this stage is to get the engine mount/king post placed perfectly, as everything else is coming off of, or going to, this central component.

Wing root fittings still need to be made and mounted (Fig. 9-24). There are two channels or fittings per side: on the king post's fore and aft base—where it rests on the fuselage tubes. Out of each forward base, there's a ¾″ tube, into which the wings' fittings of ⅝″ inserts for about 6″. This ⅝″ wing root fitting is constructed in such a fashion as to give the wings a forward sweep of 5 degrees.

On the aft base of the king post, two prongs stick out to mate with forks from the aft matching fittings off of the wings. These two prongs are 3 ¾″ apart vertically.

Connect one of these ⅝″ inserts to the wing root tube—the inboard end of the 3 ½″ tube—and the forks on the beefed-up portion of the inboard rib. These should match the forward ¾″ tubes at the king post base, and the two prongs from the aft king post base.

Before permanently attaching these rear two prongs and forks, you want to make the two compression struts, and rig the

Fig. 9-24. With the compression strut clamping wing to rudder assembly, a special chromolly rig takes its place at the king post for reinforcement. Notice wing root fitting going from wing spar tube into king post's forward base.

wings' 6-degrees dihedral (Figs. 9-25 and 9-26). The compression struts are made of 2″-diameter .058 thick-walled chromolly, and are 6′5″ long. There are flanges on both ends of each for attaching to the king post and wing saddle which sits on the main wing spar about midspan.

This wing spar saddle is different from other saddles (Fig. 9-27). It has a box or square tube coming off of it that mates with the compression strut. The saddle is bonded and riveted every 1″ with ⅛″ rivets. Drill the rivet holes in the saddle; however, don't drill the wing spar, for now.

Place the wing on saw horses so it sets at six degrees dihedral. The spar alignment fitting is then inserted into the ¾″ tube on the base of the triangle of the king post. When these are

Fig. 9-25. Compression strut and king post pinned at common mating point.

Fig. 9-26. Detail of compression strut and king post.

lined up, the wing has six degrees dihedral and five degrees forward sweep. Then place the compression strut into position.

Pin the strut at the king post and mark where it touches the wing spar, with the wing swept forward 5 degrees, and 6-degrees dihedral. It's at this joining point the saddle is attached. Apply the saddle, and, using the previously drilled rivet holes of the saddle, drill on into the wing tube. Clean the area with MEK, apply 9410 adhesive, and pop-rivet in place.

Fig. 9-27. Right wing being adjusted for compression strut/king post hookup. Notice saddle reinforcement on wing tubing, where compression strut connects.

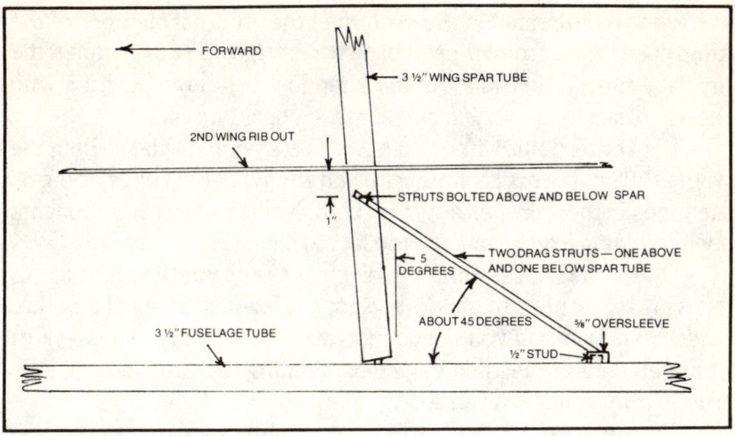

Fig. 9-28. Wing drag strut schematic.

Now you've got the wings fitted, pinned and nailed exactly on all angles. Bond and bolt the control system. The seat and other attachables in this area should be in position and tack-riveted. (Tack rivets are temporary aluminum rivets which will, after final adjustments, be drilled out and permanent, aircraft quality rivets installed.)

One of the final, and most important, adjustments to the wings are the drag struts. This drag strut helps maintain the 5-degree forward sweep of the wings, as well as their 6-degrees dihedral.

The actual struts are ½" chromolly, beat flat on one end and drilled to take a ¼" bolt. There are two per wing. On the other end, the two tubes come together at a common ⅝" over sleeve.

Weld on the drag-strut-stud (Fig. 9-28) to the engine mount, before you slide on the engine mount. This stud is the mate to the strut's ⅝" over-sleeve. Make this stud from a ½" bolt, planed off on each side. It's almost the same as the wing root fitting. You're going to have to mount the drag strut to that stud. Work on a stud-to-spar basis, so that it's clamped into position.

From the wing root of the spar tube, or the inboard rib, measure out 17". This should put you about 1" inside the second rib. At this point, drill two ¼" holes in the 3 ½" spar tubing, one above and one below. The two tubes comprising the drag strut will run from the engine mount stud to these upper and lower ¼" holes. Measure and cut the ½" chromolly tubing accordingly.

Run the push/pull rods to the ailerons and to the elevator. The aileron tubes run from the control system to the ailerons through a clevis fork arrangement. The elevator rods run from the control system to a bellcrank on the bottom of the aft crossmember of the king post, and, from there, on back to the elevator along the fuselage tube. The rod aft of the king post is held in position with the use of fair leads made from plumber's bushings.

It's at this point when you have "Joe Average" just pulling the whole P-38 out of his Lightning kit and simply sliding everything on the tubes in their proper location. If you've had secondary thoughts about building from plans, it's too late to change at this stage.

One of the final assembly projects is the cosmetics. It's back to the wings. You need to make slugs for the leading edges. These are made from foam, so you need a hot-wire for forming. These slugs fit in between the leading edge ribs, coming back to, and partially around, the wings' spar tube (Fig. 9-29)

Once you've cut these to form, stand the wing on its trailing edge on sawhorses. Then all you have to do is drop the slugs in place, bonding as you go.

Next, cut .4mm birch plywood to fit, all the way back to the wings' main spar tubes (Fig. 9-30) Then, using the same techniques as with the B-10 and U-2, bond the plywood leading edge, and use rubber bands, or surgical tubing, to strap in place until fixed.

This procedure lends itself to a heavy leading edge, but, with a craft like the P-38—or any homebuilt ultralight, for that matter—

Fig. 9-29. Skinning the leading edge. Precut foam slips down between each rib and is covered by plywood. Notice drag fittings angling to the bottom of this inboard tip, and the compression strut wing connect stub sticking from plywood covering.

Fig. 9-30. Chalk-lining the leading edge for final trim directly above the centerline of the wing tubing spar.

where it encounters incredible abuse from ground handling, it's an added attraction.

From here, it's simply a matter of covering the Lightning, putting on the engine, and going flying.

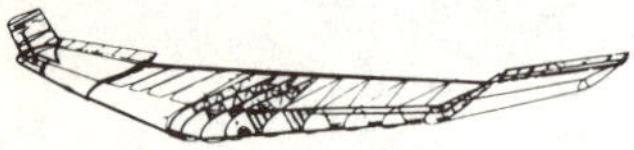

10

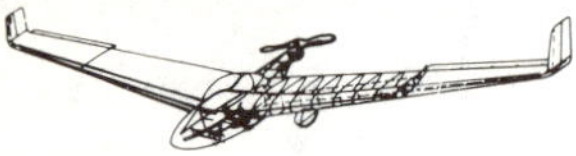

Flying the Mitchell P-38 Lightning

The P-38 Lightning, like the B-10, flies with the use of the NACA 230-15 airfoil. It has been described by experts as one of the finest wings ever developed for good lift and inherent stability (Fig. 10-1).

The Lightning's wings are swept forward 5 degrees, and carry 6 degrees of dihedral. The 230-15 has further been modified to create a 3-degree twist. This will be explained in detail in the "stall" section (Fig. 10-2).

It is an ideal situation when aircraft loading can be concentrated right on the established center-of-gravity. This is one of the reasons for the P-38's wings being swept forward, since the pilot is about the only weight carried, with the exception of a few pounds of fuel.

The craft, as a whole, may appear to be a mixture of "odds-'n-ends," but, it is, in fact, an aerodynamically efficient, fun-flying bundle of joy. It's also the first Mitchell machine to carry a tail assembly, thus breaking away from their traditional flying-concept, as exemplified with the B-10 and U-2.

The pilot sits up front in a bucket seat. Even this seat is built in such a fashion as to create a minimum contingent of drag. It's a comfortable seat, and the view is relatively unrestricted.

To the left and right, from the pilot's seat, mounted on the two fuselage tubes forward, are the engine controls and instrumentation (Fig. 10-3). These take in standard power-throttle, choke, brake, airspeed indicator, ignition switch and altimeter on the left; and rpm, engine temperature and magnetic compass on the right.

Fig. 10-1. The P-38 ready for fire-up.

The elevator/aileron control stick is located between the legs, military fashion. Slight movements of the joy stick create sufficient amounts of control response. Your feet rest forward on large, wooden rudder pedals. These pedals are not the normal push-type, but rather rock forward and backward on crossbracing. They are activated, more or less, by ankle/foot rotation.

The Lightning's front end looks to be a metallurgist's nightmare, but, in actuality, components keep with Mitchell's

Fig. 10-2. Lightning in flight. Notice forward 5-degree sweep of wings, the extra-long ailerons, and wing ribs silhouetted through fabric covering. This forward sweep is to put the pilot as close to the CG as possible.

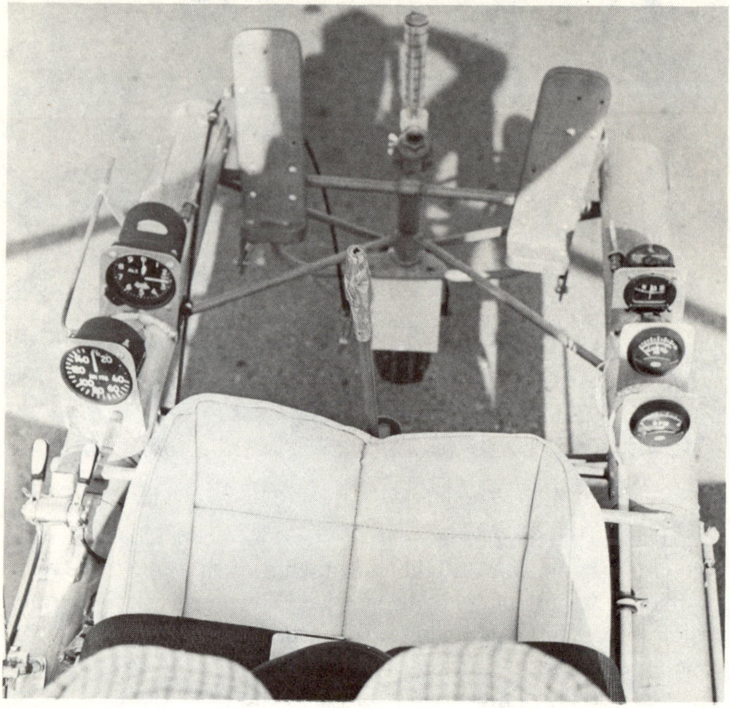

Fig. 10-3. Instruments, gauges and controls, as seen from atop the pilot's seat.

motto: "all parts are proven worthy." Take, for example, the fork holding the nosewheel. You couldn't ask for more of a proven product than a fork from a Schwinn bicycle—slightly modified, of course.

Twin vertical rudders adorn the P-38's tail section. However, they are more than mere cosmetic attachments. Besides giving the craft more working surface, it makes building the Lightning much easier for the enthusiast. Imagine the problems of running back with two fuselage tubes, then joining them into one piece, extending that aft to one rudder control surface? It was a wiser designer who decided to simply extend *both* fuselage tubes rearward, and place a rudder on *each*.

A thick, horizontal stabilator takes on the same appearance as the wings. The camber is of such thickness as to create a very efficient control surface at any speed. Like the P-38's ailerons, the elevator is a "symmetrical" airfoil, its upper and lower camber being the same.

The Lightning features full wing-spanned ailerons, set below the wing itself. Much like the stabilators on Mitchell's B-10 and U-2, these airfoils fly in their own airstream, making them extremely effective, especially during slow maneuvering.

The landing gear creates some parasitic drag, but it's kept to a minimum through selected components. Landings are buffered by ⅞" fiberglass rods on the main gear, which are tough and durable.

The king post and compression struts come together almost directly above the pilot's head. It's into the king post that most flight loads are filtered through the compression struts. This area, which also supports the engine and fuel tank, also is the site of much drag.

Assembly and Teardown

The P-38 is made to be trailered, and, to that end, it is streamlined (Fig. 10-4). A light, two-wheeler fishing boat trailer does nicely, limiting auto/trailer hitch loads to around 50 pounds.

Fig. 10-4. P-38 is trailered on a 19' bed. This can be an old boat trailer, or you can make your own. Notice compression strut holding wing flush. When trailered, the P-38 is streamlined.

Simply deck it in with ½″ wood reinforcement to carry your weight walking around.

When sitting with its wings folded, the craft is carried-faced into the wind. It's sort of tail heavy, making hitch loads even lighter. Two hinged ramps help secure the Lightning while enroute, as well as a tail tie-down.

Once at your destination, detach trailer from vehicle, and its nose will drop to the ground. From there, unlatch the ramps and swing them to the earth forward. Disconnect the tail tie-down, and walk the Lightning forward and down the ramp. It's best to walk the lightweight forward by grasping the two tail-skids, letting its weight and gravity work for you (Fig. 10-5).

On each wing, there's a special attach fitting at the rudder that holds it (the wing) vertical—the leading edge toward the ground. There is, also, an aluminum attach bracket that mates the wing to the top of the king post. Remove these two restrainers (Fig. 10-6).

Now walk the wing away from the ship and into the horizontal flight position. As you walk it forward, rotate it from the vertical position into the flight attitude. Have some type of prop to put under the wing to hold it in approximate flight position.

Walk around to the front of the ship and release the compression strut from where it attaches to the now unfolded wing. When you remove the strut's retaining pin, let it move on its own axis up to, and over, the saddle area. The opposite end of the strut is then swung up and over—almost 180 degrees—and attached to the top of the king post.

After removing the wing support, walk back inboard and insert the two retaining pins—one forward and one aft—to hold the wing in position (Fig. 10-7).

This gives you a three-point locking system for the wing. It is locked/pinned at the top of the king post, at the forward and aft base of the king post, where the fuselage tubes run through.

Repeat this process with the opposite wing. Teardown is nothing more than a reverse of this process. One person can put it together or tear it down for trailering within five minutes.

Walkaround Inspection

As you walk up to the Lightning, inspect the seat area (Fig. 10-8). Place helmet, goggles and gloves in the seat. The pilot chair arrangement is modified to your pleasure. Therefore, you will know the points in particular to check for safety.

Fig. 10-5. P-38 as it comes off the trailer.

Move down the right-hand wing first. Inspect the compression strut attach fitting at the king post, making sure it's secured and pinned in position. Carefully look over the wing's leading edge for the usual—dents, cracks and holes. Being a homebuilt, with wings folding back for trailering, the leading edge gets a lot of handling abuse. The P-38's forward edge is strengthened for this reason.

On the outboard leading edge, there's an attach fitting, where the compression strut attaches to the wing from the king post—inspect it. Check the wing tip for fabric holes, tears and wear.

Coming down the wing's backside, check the aileron and its movement. Go over its connection hinges and push/pull rod activator. Since the aileron is constructed and flies much as the wing, check it carefully.

Fig. 10-6. Off the trailer, one wing still folded, the P-38 is readied for flight.

Next, turn your attention to the engine. Scan the engine mounts for cracks or wear and attachment firmness. Being a mid-engine above the craft, the propeller shouldn't be dented or chipped with debris picked up off the ground, but inspect it for cracks and wear. Survey the belt drive from the reduction system, and the belts themselves, for tension and signs of fraying. The tuned exhaust system is kind of sensitive. It is maintained by rubber shock absorbers, springs and wire. Eyeball the fuselage tube/king post brackets for wear. Signs of bracket friction will appear as black scrub marks on the aluminum tubing fuselage.

Go back to the tail and inspect for dents, cracks and unusual bits and pieces that could have been picked up from accidental tail

Fig. 10-7. Ready to go. The Lightning can be assembled for flight by one person within five minutes, or loaded for taking home just as fast.

touchdown. Move the rudder back and forth, listening for normal sounds of the activator cables channeling through the tubular fuselage housing. Next, look over the elevator and its fittings for motion. Examine the push/pull rod that manipulates the elevator.

Move on around the left side, inspecting it much as you did the right, i.e. aileron, wing tip, compression strut attach points, leading edge, back to the cockpit section.

Check the seat belt and shoulder harness, as you pass the seat area. Also inspect the throttle connections. Looking back on the engine from the front, examine for fuel quantity, fuel cap and strainer (Fig. 10-9). Turn around and survey the flight instruments, engine instruments, rudder pedals and accessories.

Squat down and examine the nosewheel and braking system. Make sure the left and right main wheels are in order, along with the fiberglass landing gear rods. And you're ready to fire-up.

Startup and Taxi

Even though the electric starter is an option, assume the basic model is still started by means of the "Armstrong" method. Starter mechanism is a self-retracting starter pulley, much like those used with a chain saw or lawn mower.

Reaching down on the left-hand fuselage tube, insure the ignition switch is "off." Now, close the throttle, lever the choke "on," and pull the starter rope through three times. Then, switch "on," throttle cracked, choke "off," and pull for effect. It should catch on the first turnover.

The preceding is for cold starts only. Warm, or hot, starts eliminate the need for this priming phase. Simply turn the ignition "on," open the throttle a tad, and pull it through.

Again, this is with the basic ship, with the engine and starter systems just as they were assembled from the box. Most pilots

Fig. 10-8. Three-quarter view from rear highlights fully wing-spanned ailerons, large-surfaced elevator and rudders. Walkaround inspection begins at right front.

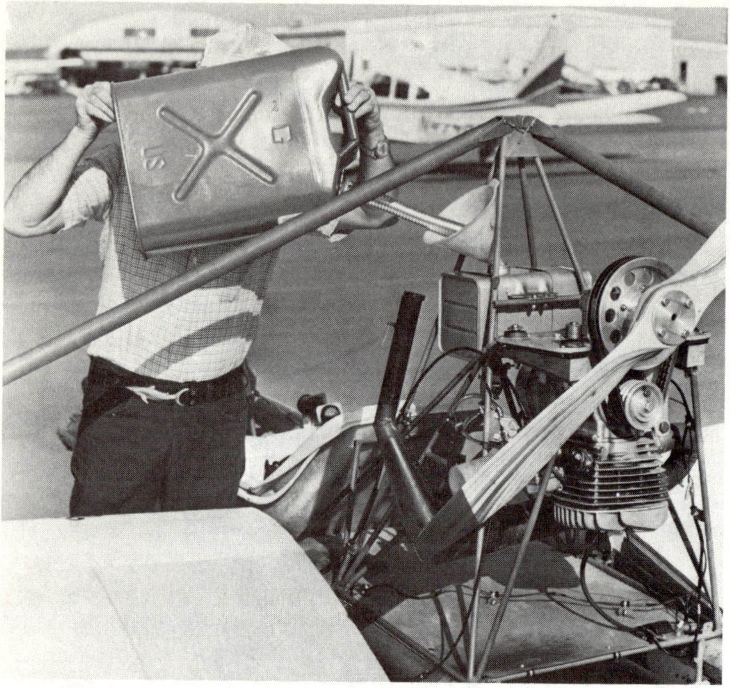

Fig. 10-9. Fueling 1½ gallon fuel tank. Long range 4-gallon tank is optionally available.

modify the self-retracting starter mechanism so as to pull it through while sitting in the cockpit strapped down.

If you haven't done so prior to engine start, don the helmet, gloves and goggles you put in the pilot's seat earlier. Get aboard, strap down, and you're ready to taxi out . . . *almost*.

After the engine catches, with the throttle cracked, it'll settle at approximately 1000 rpm. Before taxi, let the engine temperature indicator come off the peg.

Apply a little throttle and the machine moves. Taxi rpm is 2000-2500. Engine temp will stay below 300-degrees F while on the ground.

Taxiing is fairly simple, as rudder/nosewheel steering is adequately responsive. For the normal tricycle-gear pilot, though, differential braking is missed. The single nosewheel brake is sufficient, and braking is enhanced by leaning forward, thus shifting the CG closer to the braking area.

Spend a couple of hours getting the feel of the craft through low- and high-speed taxiing. Run it around the airport area, up and down the taxi strips. You may even want to set up a circuit of obstacles or a maze to run through at various speeds. Test the braking at high speeds, as well as low, in corners or on the straight. Practice the art of parking—actually pulling into and turning around in an enclosed area—as this is a skill in itself.

Get the feel, in general, of the Lightning's ground handling to the extent that you are comfortable driving the machine in all modes. It should come as second nature before attempting to proceed.

High-Speed Taxi and Ground Skimming

Once you've got the feel of the aircraft at slow to moderate speeds, take it out on the runway for some high-speed work.

Sitting on the runway, lined up and ready to go, make sure the brake lever is fully down and "off." That completed, slowly add full power. By now, you'll know how quickly the ship picks up speed.

Let it accelerate to approximately 25 mph, keeping it on the centerline, then back off the power, letting it decelerate on its own. Turn around and repeat the process back to the other end, or taxi it back for another go. Keep this up until satisfied with your competence.

Repeat the above, taking the P-38 up to 30 mph. At this speed, bring the control stick rearward slightly, making the nosewheel light, or rotating it off the ground completely (Fig. 10-10). Care should be exercised *not* to lift off.

The next phase consists of building speed to about 35 mph, rotating the nosewheel off the ground, as before. With a further nudge on the control stick rearward, lift the ship completely off the ground to a height of 3-5 feet.

Hold the machine at that altitude, staying over the runway centerline, and ground skim (Fig. 10-11). Reduce power and set it

Fig. 10-10. At about 30 mph, bring the stick back, making the nosewheel light, or rotate it off the ground completely.

Fig. 10-11. Ground-skimming height. Take it off, skim the runway, set it back down. Repeat this routine, until thoroughly familiar with rotation, liftoff, leveling, flare-out, touchdown, and braking.

back down. Repeat this procedure again and again. Get the lift-off and touchdown sensations. Feel the sponginess of the fiberglass landing gear. Experiment with full and partial braking. Take and analyze notes. Know yourself and the aircraft before proceeding.

Takeoff, Climb and Landing

Give the P-38 full power and accelerate, just as ground skimming. Once off the runway, the Lightning will climb like crazy at that power setting. So level off and let the airspeed build to 40 mph, reducing power to 6800 rpm.

Redline on the Honda 250cc is right at 7500 rpm. This power setting will net you 300-350 fpm climbout, while keeping the engine temp in the neighborhood of 450-degrees F. Temperature redline on the two-stroker is 550-degrees F.

After establishing climb attitude, climb straight out, until 500 feet of altitude is gained. Then make a shallow left or right pattern turn. Stay in the traffic pattern for a number of touch-'n-goes.

Around the airport, recommended speed is 40-45 mph, although you can carry a low 35. Coming over the fence at 35, you'll want to carry power until actual wheels-to-earth mating. The plexiglass rod landing gear cushions dropping it in, without the usual secondary rebounds. This gear, new like the whole machine, is very forgiving.

Next, make a normal 40-mph climbout, only this time leave the traffic pattern (Fig. 10-12). Make a few shallow turns left and right. You'll notice excellent response because of the full wing-spanned ailerons mentioned earlier. And, the large elevator surface gives quick and effective pitch control.

Leveling off at altitude, let the speed build to the upper 50s—Depending on variables, it may just tick past 60. An easy cruise of 53-55 is established at 6500 rpm. Expect airspeeds to range from mushiness and wallowing at about 27 mph on the bottom side to fairly zipping along right at 60 on the top. Test the different power settings, marking down speeds attained, engine temp. and altitude.

Stalls and Steep Turns

To test the P-38's turning capability, it's best to do it *gradually.* You're sitting out in the open, with an almost 360-degree wrap-around view. Going from level flight into immediate steep turns, with nothing between you and the ground way down below, can be a bit frightening at first.

Try a normal, standard turn first. A standard turn is approximately 3-degrees per second, or two minutes for 360 degrees, if you time it. There is no turn or bank indicator on the P-38, other than the magnetic compass.

Complete a number of turns left and right—until you're comfortable in both directions. Establish horizon references in both directions, using rudder pedals or your toes, whichever is higher, as reference.

Next, ease into a steep turn, usually 45 degrees of bank, or whatever bank that you consider steep. Practice these steep turns left and right until you have to increase power to hold altitude.

Remember to keep notes. This may be a bit tricky, since you're out in the open air, but the use of a knee pad/board will make this possible. Note altitudes, airspeed entry and exit, horizons, power settings—just about anything you can think of. These notes are very important, especially on post-flight analysis for the first few flights.

Fig. 10-12. Get a comfortable feel of the Lightning in shallow-medium turns. It's here that you'll be doing most of your fun flying.

Fig. 10-13. Speed around the traffic pattern is 45-50 mph, although it's only a recommendation.

Stalls come in a completely different category. You'll find that there's not the same characteristics that most pilots are accustomed to. The nose and wings don't "break" at the burbling point.

Phase into stalls through slow flight. In a straight and level attitude, ease the power back, while maintaining altitude. You'll notice that about 30 mph, or a little above, the Lightning gets mushy and sluggish. Although you still have complete control, and the aircraft systems are responding adequately, you can feel—seat of the pants—and observe it react slower than normal.

This mushiness will increase until, at 27 mph, it is wallowing heavily. Below this airspeed, the machine is going to lose altitude in the traditional "falling leaf" manner.

Recovery is standard—nose, power, wings (rudder). Almost from the time you add power, the little craft perks up. Like the U-2, there's a 3-degree twist in the P-38's wings, allowing the inboard section to stall, while the outboard is still flying. Checking the altimeter for altitude loss, its needle will not have moved much, if any, it's so minimal.

The twin vertical rudders and free-flying ailerons keep you in positive control of the Lightning throughout slow flight and ensuing stall.

Descents and Landing

Unlike the B-10 and U-2, the P-38 has a very *low* glide ratio. Once you cut the power on the Lightning, it begins a marked descent. It's not made for soaring or gliding, as are the B-10 and U-2. However, this characteristic can be used to work favorably for you.

Descent and approach to landing speeds are, basically, your choice (Fig. 10-13). Super-quick drops from altitude are accomplished by maintaining 55 mph, with the throttle closed. Power is carried right until touchdown for previously explained reasons — large amounts of parasitic drag. Once you've chopped the power, be ready to put the nose down, because the speed bleeds off rather rapidly.

Once on the ground, it's just a matter of closing the throttle and either applying braking or letting it roll out. Taxi back for another go at it, or put it on the trailer, and head for home.

Weights

Even though the Mitchell P-38 Lightning is very dirty, in the sense of creating lots of parasitic drag, it still carries performance figures among the top few craft in the ultralight field (Fig. 10-14).

What gives it such good specs is that except for a few minor parts, it's an all-aluminum, tubular structure, carrying a 200-pound empty weight. Max gross comes in right at 450, giving you a 250 pound usable limit.

Since there's no baggage to speak of, and the 1½ gallon fuel/oil mix comes in right at 10 pounds, it leaves you with 240 pounds to work with.

The craft, as is, will push a max pilot weight of 230 pounds in the mid-50 mph range. Pilots heavier can be carried, however, performance is greatly sacrificed.

Optional equipment for the P-38 includes a wrap-around cabin, electric starter, running lights for night ops, floats for water

Fig. 10-14. With pilot hanging out in the breeze, the P-38 is aerodynamically "dirty."

Table 10-1. Mitchell P-38 Lightning Performance and Specifications.

POWER

Engine Type: Honda
Number of Cylinders: 1
Horsepower: 20@7500 rpm
Propeller: Wood—Hand carved
Propeller Diameter: 50″
Propeller Type: Fixed pitch
Aviation Fuel Grade: Regular auto fuel/oil mix
Power Loading: 22.5 lbs/hp

SPECIFICATIONS

Wingspan: 26 ft
Wing Area: 110 sq. ft.
Length: 17 ft
Height: 5 ft
Seating Capacity: 1
Maximum Gross Weight: 450 lbs
Empty Weight: 200 lbs
Usable Weight: 250 lbs
Fuel Capacity: Standard Tank — 1.½ gal
Long Range Tank — 4 gal

PERFORMANCE

Cruise: 55 mph
Stall: 27 mph
Takeoff Distance: 200 ft
Landing Distance: 95 ft
Best Rate of Climb: 350 fpm
Service Ceiling: 9000 ft
Fuel Consumption: 1 ¼ gph

lovers, skis for cold-weather buffs and/or 20″ rough-terrain wheels for those who prefer the outback. These add-ons will, naturally, add some weight to the machine, but, in most instances, will be offset by streamlining, in one form or another.

Looking at it in another way — If the ship, being this dirty, still produces top-of-the-line figures for the ultralight class, it can only do one thing as it's cleaned up — improve performance-wise, and this has a lot to do with working with present weights.

Homebuilt Flight Testing

11

Testing any new homebuilt can be either a pleasant and rewarding experience or one of actual *terror*. Now that you, as the *builder*, have put your Mitchell Wing together, you, as pilot, must test it, both on the ground and in the air (Fig. 11-1).

It's important, even before beginning any type of test program, to have an in-depth knowledge of all Mitchell Wing components and their functions. If it's there, it's there for a purpose. Know that purpose and ask yourself: *do you think it will perform that function in a safe manner*?

As you approach the first test, begin systematically going through the Wing, system by system, confirming that it's operating properly. Tape weights to the stabilators, and have someone hold the wingtip rudders, while you work the controls from the cockpit. This is done to detect any excess play or weakness in the system.

Next comes extensive slow-speed taxi tests. Here you get the feel of nosewheel/rudder response and coordination. Set up a course or taxi around the airport. Engine and engine instrument performance are monitored, as well as brakes and simple steering and handling patterns. Watch for vibration, shimmy, brake shudder and imbalance of tires.

Taxi Testing

Graduate onto the runway and high-speed taxi testing. *Caution:* High-speed taxiing is accomplished well below lift-off speed. Here you're testing the machine for steering, tracking and brake effectiveness, not necessarily for flight controls.

Fig. 11-1. B-10 and U-2

Once you've tightened up the Wing and made adjustments, it's time to plan for its flight testing. I say *plan* because that's just what you do —*plan* a sequence to be followed, and schedule a day so that a "go" or "no-go" decision depends on weather only. This day should be clear, with no or very little wind, and be an early morning flight, if possible.

Plan the flight from an out-of-the-way airport with a 4000-5000-foot runway. Have only those people around who are essential to the first flight, including one hand-picked assistant (Fig. 11-2). Provide the assistant with some type of vehicle and a list of phone numbers to the nearest ambulance, sheriff's, and fire department's, if there's none located on the field.

By way of personal equipment, you're going to want a crash helmet, flying suit or equivalent, jump boots or heavy shoes and gloves. Also, a large knife, fire extinguisher and parachute should be present but are optional (Fig. 11-3).

Clothing and emergency gear may change later, but for flight testing these are basics. Don't let all of this scare you — these are just normal precautions. Chances are greatly in favor of your never having to use them.

Skimming

If the weather and ship criteria are met, start up and taxi onto the main runway. Now, almost like the high-speed taxi tests, add full power. Let the bird accelerate to liftoff speed. Instead of shutting down, as done previously, ease the stick back until ground separation occurs — 5-10 feet. Hold it there momentarily, then chop the power, flare out and land. Taxi back, and repeat until familiar and comfortable with taking off, ground skimming, and landing.

Bring the craft back in, and give it a good going over before proceeding with the next phase. By now, you'll be excited and anxious to get on with some *real* flying. *Control the urge*. You're almost there but not quite.

What you're accomplishing with the constant going over of your machine and making adjustments is fine tuning it — or, as they say, "working the bugs out." When you've tightened it up and made those adjustments, it's on to the next phase.

Flight

Back out on the runway again, you give it full power, let it accelerate, lift off, but instead of chopping the power, you let it continue to climb out straight ahead. Fly the pattern, always keeping within gliding distance of the runway, and make a number of "touch-and-gos." Repeat until you're familiar and comfortable flying the airport traffic pattern. The only thing new you've added is the climbout and basic turns. Again, it's back in for another checkover.

For the next phase, make a normal takeoff and departure, using a fairly flat climb angle. Take the wing up to 3000-4000 feet, staying right over the airport at all times. While climbing out, keep an eye on engine or cylinder-head temperature. Also, on the way up, check general handling characteristics — bank and pitch.

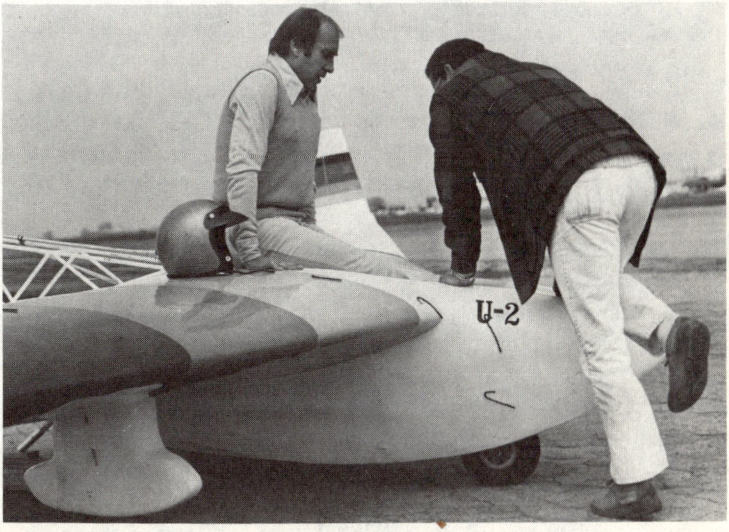

Fig. 11-2. Ready to test. Have a trusted assistant at hand.

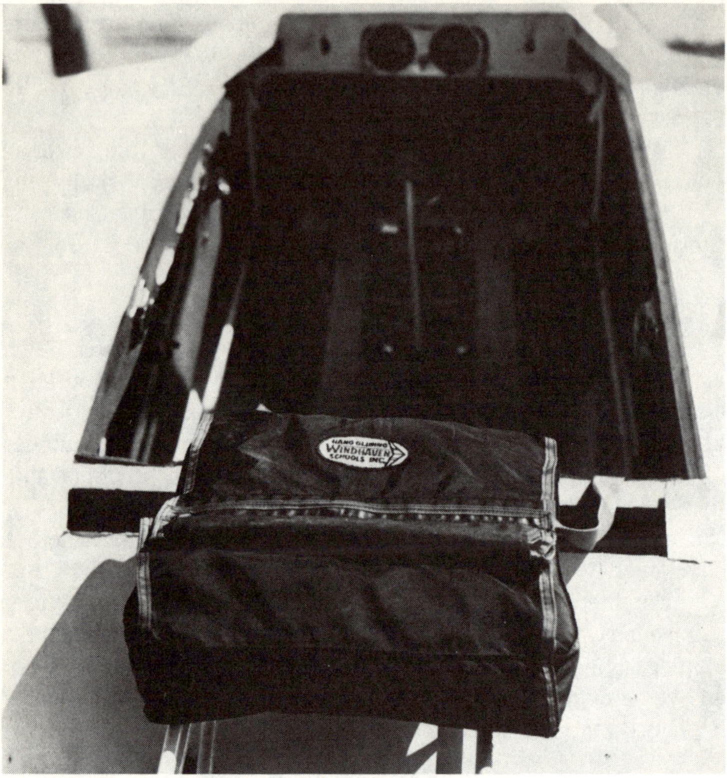

Fig. 11-3. Parachute hooked to aircraft's main structure. Simply toss it out to save your ship and yourself.

Once at altitude, start the program by rolling from bank to bank. This will quickly give you the feel of the controls and relationship between wingtip rudders and the stabilator. Begin with 180-degree turns in both directions, and work up to 360s likewise.

At this point, gradually increase the back pressure, until you get a stall. Check to see how violent the stall, and how quickly it flies again when forward pressure is applied to the stick. Try stalls with and without power, and in and out of turns. If you've brought a knee pad along, jot down flight characteristics and speeds. Now throw in some slow flight, and call it quits. Chop the power, glide down and enter the traffic pattern.

Follow your flight plan, and for the first landing from in-depth testing, use a standard pattern with some power, clearing the engine every now and then. Use 1.4 to 1.5 times the newly determined stall speed reading for approach. Also, figure on about 1.6 times stall speed as giving the machine its max rate of climb. These figures may change as your skill on, and knowledge of, the machine grows. But, for now, these are safe speeds.

It should be emphasized that anytime you receive unusual control responses, break off testing immediately and land until a valid reason is obtained and corrective measures taken. If necessary, modify and/or obtain help from a more experienced pilot and/or builder.

After the flight, make an immediate expansion of your flight notes or remembrances, either on paper or with a tape recorder — then make a detailed inspection of the aircraft and initiate adjustments as needed.

Keep your first flight to 30 minutes or less. There are many items to check before your next flight. After a thorough check of the engine, controls, flight surfaces and such, proceed with scheduled and preplanned flight testing. Save takeoff- and landing-roll distances and distance to clear a 50-foot obstacle until last.

Spin characteristics of a homebuilt are likely to be unique to each one and difficult to predict. Learning spin conditions and recovery shouldn't be undertaken, unless, of course, you're an aerobatic student. It isn't quite clear that spinning a newly constructed Homebuilt (unless designed for aerobatics) is useful. Ultralights should *never* be spun, as a general rule. Infinitely more useful is understanding the flight feeling just before a stall, both straight ahead and banked flight, since this is the deadly margin of flight which can occur at low altitudes in the approach pattern, and serve as a warning prior to entering a spin. Standard spin recovery is used when, and if, a spin situation is encountered.

Enhancing any flight test on any homebuilt is the camaraderie and association with others with the same craft. Talk to them and listen to what they have to say. Remember, they've gone through the same thing, and will give you some idea of what to expect.

Flight Test Checklist

No test guide or check sheet can completely take the place of your mind and memory, but are only used in conjunction with them as a guide to logical order and as a stimulation. They should work hand-in-hand to gather facts.

These facts should be expanded, either on paper or with the use of a tape recorder, as soon as possible following each flight. The following is only an example, and can be reworked to suit your need:

General

Date:
Flight Number:
Aircraft Weight As Flown:
Flight Test Forward CG:
Flight Test Rearward CG:
Object of Flight:

Taxiing

Right Rudder Response:
Left Rudder Response:
Brake Response:
Shock Response:
Engine Temp:

Pre-Flight

Throttle Medium High:
Throttle High:
Tuned Exhaust:
Engine Temp:
Tachometer:
Altimeter:
Trim:
Canopy Latches:
Seat- and Shoulder-Harness:
Stick Forces:

Takeoff and Climb

Controllability:
Stability:
Balance:
Stick Forces:
Indicated Airspeed:
Engine rpm:
Engine Temp:
Rate of Climb (fpm):
Takeoff Distance:

Normal Cruise

Indicated Airspeed:
Engine rpm:
Pressure Altitude:
Outside Air Temp:
Stability:
Controllability:

Stick Forces:
Engine Temp:

Top Speed — Level Flight

Indicated Airspeed:
Engine rpm:
Pressure Altitude:
Outside Air Temp:
Stability:
Controllability:
Stick Forces:
Engine Temp:

Stall — Power On

Indicated Airspeed:
Indicated Airspeed Gear Down:
Engine rpm:
Stall Warning:
Recovery Altitude:
Recovery Movements:

Stall — Power Off

Indicated Airspeed:
Indicated Airspeed Gear Down:
Engine rpm:
Stall Warning:
Recovery Altitude:
Recovery Movements:

Glides and Descents

Max. Glide Angle (mph):
Min. Altitude Loss (mph):
Max. Descent Speed (mph):
Normal Descent Speed (mph):
Flutter or Vibration:
Stick Forces:
Controllability:
Stability:
Engine Temp:

Approach and Landing

Approach Speed Power On:
Approach Speed Power Off:
Flare Out Speed:
Touchdown Speed:
Gear Operation Speeds:

Gear Emergency Operation:
Landing Distance:

Remarks

1. ____________________
2. ____________________
3. ____________________
4. ____________________
5. ____________________

Index